A Student's Guide to Infinite S___ ___equences

Why study infinite series? Not all mathematical problems can be solved exactly or have a solution that can be expressed in terms of a known function. In such cases, it is common practice to use an infinite series expansion to approximate or represent a solution. This informal introduction for undergraduate students explores the numerous uses of infinite series and sequences in engineering and the physical sciences. The material has been carefully selected to help the reader develop the techniques needed to confidently utilize infinite series. The book begins with infinite series and sequences before moving onto power series, complex infinite series, and finally Fourier, Legendre, and Fourier-Bessel series. With a focus on practical applications, the book demonstrates that infinite series are more than an academic exercise and helps students to conceptualize the theory with real-world examples and to build their skill set in this area.

BERNHARD W. BACH JR. is the Director of Undergraduate Laboratories at the University of Nevada, Reno. His research interests focus on gamma ray, X-ray and UV spectroscopy, and the manufacture of diffractive optics and spectroscopic instrumentation. He has contributed to the design and construction of scientific instruments for numerous space-flight missions and synchrotron light sources around the world.

Other Books in the Student Guide Series

A Student's Guide to Infinite Series and Sequences

BERNHARD W. BACH JR.
University of Nevada, Reno

CAMBRIDGE
UNIVERSITY PRESS

CAMBRIDGE
UNIVERSITY PRESS

University Printing House, Cambridge CB2 8BS, United Kingdom

One Liberty Plaza, 20th Floor, New York, NY 10006, USA

477 Williamstown Road, Port Melbourne, VIC 3207, Australia

314-321, 3rd Floor, Plot 3, Splendor Forum, Jasola District Centre, New Delhi - 110025, India

79 Anson Road, #06-04/06, Singapore 079906

Cambridge University Press is part of the University of Cambridge.

It furthers the University's mission by disseminating knowledge in the pursuit of education, learning and research at the highest international levels of excellence.

www.cambridge.org
Information on this title: www.cambridge.org/9781107059825
DOI: 10.1017/9781107446588

First published 2018

A catalogue record for this publication is available from the British Library

Library of Congress Cataloging in Publication data
Names: Bach, Bernhard W., Jr., 1965– author.
Title: A student's guide to infinite series and sequences / Bernhard W. Bach, Jr. (University of Nevada, Reno).
Other titles: Infinite series and sequences
Description: Cambridge : Cambridge University Press, 2018. | Includes bibliographical references and index.
Identifiers: LCCN 2017061456 | ISBN 9781107059825 (alk. paper)
Subjects: LCSH: Series, Infinite. | Calculus.
Classification: LCC QA295 .B2245 2018 | DDC 515/.243–dc23
LC record available at https://lccn.loc.gov/2017061456

ISBN 978-1-107-05982-5 Hardback
ISBN 978-1-107-64048-1 Paperback

To Angie, Tim, Brian, Aaron, and Cam,

for their advice and support along the way.

Contents

Preface

Why study infinite series? Not all mathematical problems can be solved exactly or have a solution that can be expressed in terms of a known function. In such cases, it is common practice to use an infinite series expansion to approximate or represent a solution. For example, many differential equations have solutions that cannot be expressed in terms of known or elementary functions, yet their solutions can be written out as infinite series of terms.

Infinite series are also used to approximate the numerical values of specific functions or integrals. For example, the value of a transcendental function can be calculated using an algorithm in which the transcendental is represented as an infinite series of terms. Representing a function as an infinite series is a technique that is widely used in the sciences and engineering. The ability to expand a function as an infinite series along with proficiency in manipulating such a series is a useful skill set for those pursuing careers in the sciences or engineering.

Most of us are introduced to infinite series in a second-semester calculus course, where the material is usually treated as cursory, leaving us with only a vague sense of infinite series. Furthermore, this treatment gives little indication of the practical applications of infinite series. While the present text is certainly incomplete, the material it contains has been selected to help the reader develop the skills needed to confidently create and manipulate infinite series as well as appreciate their wide range of applications in science and engineering.

1

Infinite Sequences

1.1 Introduction to Sequences

Sequences are useful in the analysis of structures and patterns that occur in a variety of contexts and across a broad range of disciplines. Sequences occur in mathematics, biology, chemistry, and physics as well as in finance, manufacturing, and computer science. A sequence can be used to represent a mathematical structure, a manufacturing process, or the pattern of nucleotides in a DNA molecule. Sequences can also express rules of thumb or general properties of a system. For example, the f-stops 1, 1.4, 2, 2.8, 4, 5.6, 8, 11, and 16 found on the aperture ring of a camera lens essentially form a geometric sequence. This sequence represents the amount of light reaching the camera's film or sensor per unit area. Another example of a practical sequence is the Mariner's Rule of Twelfths – 1, 2, 3, 3, 2, 1 – which is a rule of thumb for estimating water depth when navigating or anchoring a ship in shallow water. Simply stated, a **sequence** is a list of **elements** or **terms** in a particular **order**. The **elements** of a sequence can be numbers, functions, names, letters, and so forth.

What "a particular order" means is probably best demonstrated by way of a simple example. Consider the sequences (A, M, Y) and (M, A, Y). While the two sequences contain the same elements, they are considered to be different sequences, or not **equal**, because the ordering differs.

1.2 Notation

There are a number of ways to represent a sequence. The notation chosen depends on the form of the sequence and what you know about it. One method is to simply **list** the elements of the sequence:

1, 1.4, 2, 2.8, 4, 5.6, 8, 11, 16	(photographic f-stops)
1, 2, 3, 3, 2, 1	(Rule of Twelfths)
$1, -1, 1, -1, \ldots$	(alternating sequence)
2, 4, 6, 8, 10, ...	(positive even integers)
3, 5, 7, 11, 13, ...	(prime numbers)
1, 1, 2, 3, 5, 8, ...	(Fibonacci sequence)

Since the sequence of photographic f-stops and the Rule of Twelfths are examples of **finite sequences**, in that they only contain a finite number of terms. The remaining sequences are examples of **infinite sequences**, where the three little dots at the end indicate that the sequence continues forever.

A sequence may be represented using **index notation**,

$$a_1, \ a_2, \ a_3, \ldots, a_n, \ldots \tag{1.1}$$

where a_1 is referred to as the **first term** of the sequence, a_2 the **second term,** a_3 the **third term**, and a_n the **nth term** or the **general term** of the sequence. The **index n** indicates the position of the term in the sequence and is typically taken from the set of natural numbers $\{1, 2, 3, 4, \ldots\}$. Index notation is useful when you recognize the pattern or rule generating the sequence, so that the nth or general term of the sequence can be expressed as a formula or a function. Using index notation, the sequence of photographic f-stops, the alternating sequence, and the sequence of positive even integers can be written as follows:

$1, 1.4, 2, \ldots, (\sqrt{2})^8$	(photographic f-stops)
$1, -1, \ldots, (-1)^{n+1}, \ldots$	(alternating sequence)
$2, 4, 6, 8, \ldots, 2n, \ldots$	(positive even integers).

Sequences may also be represented by the notation $\{a_n\}$, where a_n is the nth or general term and it is understood that the index n runs from 1 to ∞ – for example,

$\{(-1)^{n+1}\}$	(alternating sequence)
$\{2n\}$	(positive even integers),

Since the sequence of photographic f-stops is an example of a finite sequence, it is necessary to denote the end of this sequence. In this case it is customary to represent the sequence as:

$$\left\{(\sqrt{2})^n\right\}_{n=0}^{8} \qquad \text{(photographic f-stops)}$$

In this example, a subscript and superscript are used to define the beginning and end of the finite sequence: the indexing begins with $n = 0$ and ends with $n = 8$.

As this example shows, indexing does not necessarily have to begin with the number 1; it can begin and end with any possible integer. The indexing in this example was chosen in order to simplify the expression for the nth term.

It may not always be simple or even possible to find a formula for the general or nth term of a sequences. In such cases, the sequence cannot be represented using index notation, and the elements of the sequence must instead be listed to indicate the sequence. For example, note that the Rule of Twelfths, and the sequence of prime numbers are expressed as a list of elements:

$$1, 2, 3, 3, 2, 1 \qquad \text{(Rule of Twelfths)}$$
$$1, 1, 2, 3, 5, 8, \ldots \qquad \text{(Fibonacci sequence)}$$
$$3, 5, 7, 11, 13, \ldots \qquad \text{(prime numbers)}$$

The Fibonacci sequence is an example of a sequence that can be defined **recursively**. A **recurrence relation** is an expression that relates the nth element of a sequence to a previous element or elements. The **recurrence relation** for the Fibonacci sequence 1, 1, 2, 3, 5, 8, 13, ... is expressed as follows:

$$a_1 = a_2 = 1;$$
$$a_{n+2} = a_n + a_{n+1}.$$

In Chapter 4, we will develop the necessary tools to find a general expression for the terms of the Fibonacci sequence, and we will then specify the sequence using index notation.

The sequence of prime numbers 2, 3, 5, 7, 11, ... is intriguing because there is no known formula capable of generating all the prime numbers. Therefore, we are reduced to presenting the prime numbers as a list. The distribution of prime numbers is currently an open question in mathematics for which there is a related prize, the Clay Mathematics Institute Millennium Prize.

1.3 Example Sequences

Arithmetic, harmonic, and geometric sequences are three types of sequences that are easily defined because there is a constant relation between consecutive terms.

1.3.1 Arithmetic Sequences

An **arithmetic sequence** is a sequence in which consecutive terms differ by a constant amount called the **common difference**. If the first term of the sequence is **a** and the common difference is **d,** then the arithmetic sequence is represented by

$$a, \ a + d, \ a + 2d, \ 1 + 3d, \ ..., \ a + (n - 1)d, \ ... \qquad (1.2)$$

Each term of the sequence can be obtained by adding the common difference d to the previous term – for example,

$$1, \ 2, \ 3, \ 4, ...$$
$$3, \ 7, \ 11, \ 15, ...$$
$$10, \ 7, \ 4, \ 1, -2, -5, ...$$

are arithmetic sequences with the common differences 1, 4, and −3, respectively.

1.3.2 Harmonic Sequences

The terms of a **harmonic sequence** are the reciprocals of the terms of an arithmetic sequence, so a harmonic sequence is expressed as

$$\frac{1}{a}, \ \frac{1}{a + d}, \ \frac{1}{a + 2d}, \ \frac{1}{a + 3d}, \ ..., \frac{1}{a + (n - 1)d}, \ ... \qquad (1.3)$$

Using the arithmetic sequences given above, we can form the corresponding harmonic sequences:

$$1, \ \frac{1}{2}, \ \frac{1}{3}, \ \frac{1}{4}, ...$$
$$\frac{1}{3}, \ \frac{1}{7}, \ \frac{1}{11}, \ \frac{1}{15}, ...$$
$$\frac{1}{10}, \ \frac{1}{7}, \ \frac{1}{4}, \ 1, \ \frac{1}{2}, \ \frac{1}{15}, ...$$

A physical example of the reciprocal relationship between arithmetic and harmonic sequences is the reciprocal relationship between wavelength and frequency:

$$\lambda \propto \frac{1}{f}.$$

As an example, consider the strings of a musical instrument. As the strings are fixed at both ends, the longest standing wave, or the fundamental mode supported by such a vibrating string, has a wavelength λ that is twice the length of the string. This fundamental wavelength consists of a round trip along the string, with a half-cycle fitting between the nodes at the ends of the string. The other vibrational modes (or harmonics) supported by the string occur at $\lambda/2$,

$\lambda/3$, $\lambda/4$, ... When expressed in terms of the wavelength, the vibrational modes supported by the string form the following sequence:

$$\lambda, \; \frac{\lambda}{2}, \; \frac{\lambda}{3}, \; \frac{\lambda}{4}, \ldots, \; \frac{\lambda}{n}, \ldots,$$

which by our definition is a harmonic sequence in which the common difference d is 1. We could also choose to characterize the vibrational modes of the string in terms of their frequency rather than their wavelength. If the fundamental mode of the string vibrates with frequency f, then the higher harmonic modes are found at the frequencies $2f$, $3f$, $4f$, ..., which form the arithmetic sequence

$$f, 2f, 3f, 4f, \ldots, nf, \ldots$$

Note that the reciprocal of each term nf of the arithmetic sequence is $1/nf$, which can be rewritten as λ/n using the inverse relationship ($\lambda \propto 1/f$) between wavelength and frequency, thereby demonstrating that the terms of a harmonic sequence are the reciprocals of the terms of an arithmetic series.

1.3.3 Geometric Sequences

Geometric sequences occur in many different contexts and appear in problems involving growth or decay. In biology, this may be the growth or decay of a population; in physics, it may be the change in the number of particles due to a chain reaction or decay; and in finance, it may be the change in the value of an account due to interest.

A **geometric sequence** is a sequence in which there is a constant ratio between consecutive terms; this ratio is referred to as the **common ratio**. Each term of the sequence can be obtained by multiplying the previous term by the common ratio. If the first term of the geometric sequence is $\boldsymbol{a}$ and the common ratio is $\boldsymbol{r}$, then the geometric sequence is written as

$$a, \; ar, ar^2, ar^3, \ldots, ar^{n-1}, \ldots \tag{1.4}$$

– for example,

$$1, 1.4, 2, \ldots, (\sqrt{2})^n, \ldots \qquad \text{(the photographic f-stops)}$$
$$6, -3, \frac{3}{2}, \frac{-3}{4}, \ldots$$
$$2, 6, 18, 54, \ldots$$

are geometric sequences with common ratios of $\sqrt{2}$, $-1/2$, and 3, respectively. The common ratio of a geometric sequence is found by taking the ratio of consecutive terms:

$$r = \frac{a_2}{a_1} = \frac{a_3}{a_2} = \ldots = \frac{a_{n+1}}{a_n} = \ldots \qquad (1.5)$$

A simple physical example of a geometric sequence is the decreasing height of successive bounces of a ball. Consider an experiment in which a ball is dropped onto a hard surface [1]. The ball is initially dropped from the height $H_0 = 40.5$ cm, and the maximum height of each successive bounce is recorded: $H_0 = 40.5$ cm, $H_1 = 37.0$ cm, $H_2 = 34.5$ cm, $H_3 = 32.3$ cm, $H_4 = 30.24$ cm, $H_5 = 28.2$ cm, and $H_6 = 26.4$ cm. The heights of successive bounces thus form the following sequence:

$$40.5, \ 37.0, \ 34.5, \ 32.3, \ 30.2, \ 28.2, \ 26.4.$$

To recognize that this sequence of successive heights is a geometric sequence, you need to recognize that there is a **common ratio** between consecutive terms. Recall that for a geometric sequence, the common ratio between consecutive terms is given by

$$r = \frac{a_{n+1}}{a_n}.$$

Using the ratios of consecutive heights, we would find that

$$\frac{H_1}{H_0} = \frac{H_2}{H_1} = \frac{H_3}{H_2} = \frac{H_4}{H_3} = \frac{H_5}{H_4} = \frac{H_6}{H_5} = \text{constant} = r.$$

A few strokes on a calculator will confirm that a common ratio exists: $r \approx 0.9$ (to the first decimal place). Therefore, the nth or general term of the sequence of bounce heights is given by

$$H_n = H_0(0.9)^n.$$

Another example of a geometric sequence is as follows. A chemistry instructor once told me that 99% of an unwanted solution can be rinsed from a container if the container is filled with water and emptied out three times in a row. Is this statement reasonable? If we let a represent the original amount of unwanted solution in the container and let r represent the percentage of fluid that is retained in the container when it is emptied out (i.e., the percentage of fluid that is left clinging to the interior of the container), we find that

$ar^0 = a = $ original amount of unwanted solution,
$ar = $ amount of solution remaining in container after filling with water and emptying,
$ar^2 = $ amount of solution remaining in container after second rinsing, and
$ar^3 = $ amount of solution remaining in container after third rinsing.

Notice that the consecutive terms a, ar, ar^2, ar^3 form a geometric sequence. If the container is to be 99% clean, as claimed by the chemistry instructor, this implies that only 1% of the original solution remains in the container after three rinses. Expressed in terms of the geometric progression, the statement implies that

$$\frac{ar^3}{a} = 1\% = 0.01.$$

Solving for r, we find that

$$r^3 = 0.01$$

and thus that

$$r = 0.215 = 21.5\%.$$

So even if the container isn't completely emptied out during each rinse but retains some amount of the unwanted solution (up to 21.5% in this case), the amount of unwanted fluid remaining in the container is given by

$$a_n = a_0(0.215)^n,$$

and the amount of unwanted solution that has been removed will approach 99% after three rinses.

1.4 Limits and Convergence

If the consecutive terms of a sequence approach a constant or limiting value, the sequence is said to **converge** (or be **convergent**). If a sequence does not converge, then it is said to **diverge** (or be **divergent**). In many applications, it is necessary to determine whether a particular infinite sequence is convergent or divergent. It may also be necessary to determine the limiting value of a convergent sequence. In this section, we will develop the concept of convergence and introduce methods for identifying convergence and for determining the limiting value of a convergent sequence.

Put simply, a convergent sequence is one in which the consecutive elements of the sequence get arbitrarily close to some value. For example, the terms of the sequence

$$0.9, 0.99, 0.999, 0.9999, \ldots$$

can be observed to approach one (the limit of the sequence), so this sequence is convergent. More formally, an infinite sequence has a **limit** if the nth or general term a_n converges to some constant L as n becomes very large:

$$\lim_{n \to \infty} a_n = L.$$

If L is a real number, the sequence is said to **converge to** L.

If the successive terms of a sequence do not approach a limit, the sequence is divergent. A straightforward example of a divergent sequence would be a sequence whose nth term becomes arbitrarily large in magnitude as n approaches infinity:

$$\lim_{n \to \infty} a_n = \pm\infty.$$

For example, the sequence of positive integers

$$1, \ 2, \ 3, \ 4...$$

diverges. A more subtle example of divergence appears in the infinite sequence

$$\{1, -1, 1, -1, 1...\}.$$

The general expression for the sequence is $a_n = (-1)^n$. Taking the limit of the general expression as n gets large, we find that the nth term does not approach a constant value; rather, the sequence oscillates between positive and negative 1. By definition, a convergent sequence has only one limit, but the sequence $\{1, -1, 1, -1, 1 \ldots\}$ does not approach a single limit and is therefore divergent. We will revisit this example momentarily, after we develop a more precise definition of convergence.

The more precise, definition of convergence is as follows: an infinite sequence $\{a_n\}$ converges to a limit L if for every $\varepsilon > 0$, no matter how small, there exists a positive number $N > 0$ such that for all $n > N$, a_n remains arbitrarily close to L (i.e., $|a_n - L| < \varepsilon$). If the limit L does not exist, then the sequence diverges. Figure 1.1 illustrates this definition of convergence.

For a divergent sequence, the condition $|a_n - L| < \varepsilon$ cannot be fulfilled, even for very large n. Consider the previous example: $\{1, -1, 1, -1, \ldots\}$. For very large n, the general term a_n is ± 1, and the condition $|\pm 1 - L| < \varepsilon$ cannot be fulfilled for an arbitrarily small ε. Because the consecutive terms of the sequence do not approach a single value, a limit does not exist, and the sequence is divergent.

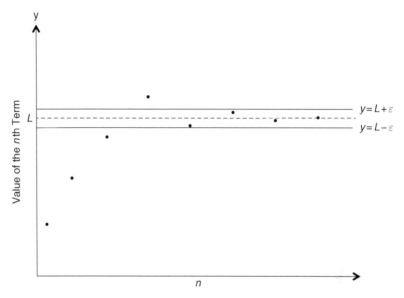

Figure 1.1 Illustration of the convergence of a sequence to the limit L.

To establish the convergence of a sequence, we need to be able to prove convergence by either resorting to the definition of convergence or being able to take the limit of the nth term or general term of the sequence. In many cases, taking the limit of the nth term will be a straightforward process. For example, consider the harmonic and arithmetic sequences representing the vibrational modes of a fixed string. Recall that we developed two sequences in Sections 1.3.1 and 1.3.2:

$$\lambda, \frac{\lambda}{2}, \frac{\lambda}{3}, \frac{\lambda}{4}, \ldots, \frac{\lambda}{n}, \ldots = \left\{\frac{\lambda}{n}\right\} \qquad \text{(sequence of allowed wavelengths)}$$

and

$$f, 2f, 3f, 4f, \ldots nf, \ldots = \{nf\} \qquad \text{(sequence of allowed frequencies)}.$$

Taking the limit of the nth term of the harmonic sequence of allowed wavelengths, we find that

$$\lim_{n\to\infty} \frac{\lambda}{n} \to \frac{\lambda}{\infty} \to 0.$$

So the sequence of allowed wavelengths is convergent, and it converges to zero. Taking the limit of the general term of the sequence of allowed frequencies, we find that

$$\lim_{n\to\infty} nf \to \infty f \to \infty,$$

and therefore the sequence diverges. In both of these examples, the limit was very easy to evaluate, as no algebraic manipulation was required. Let us now consider a slightly more complicated example, namely, the infinite sequence

$$\left\{ \frac{5n^3}{n^3 + 3\sqrt{4 + n^6}} \right\}.$$

To test this sequence for convergence, we simply need to take the limit of the general term:

$$\lim_{n\to\infty} \frac{5n^3}{n^3 + 3\sqrt{4 + n^6}}.$$

Dividing the numerator and denominator by n^3 and taking the limit, we find that

$$\lim_{n\to\infty} \frac{5}{1 + 3\sqrt{\frac{4}{n^6} + 1}} \to \frac{5}{1 + 3} = \frac{5}{4},$$

and so the sequence converges to the limit 5/4.

Occasionally, there are situations where it may be necessary to resort to the definition of convergence in order to establish the convergence of a sequence. As an example, we will use the definition of convergence to prove that the geometric sequence $\{r^n\}$ converges to the limit for constant r, where $|r| < 1$ [2]. Using the definition of convergence, we need to show that for every $\varepsilon > 0$, no matter how small, there exists a positive number $N > 0$ such that for all $n > N$, r^n remains arbitrarily close to zero (i.e., $|r^n - 0| < \varepsilon$). For $r = 0$, it is obvious that the sequence $\{r^n\}$ is constant $(0, 0, 0, \ldots)$; the limit is therefore 0, and the sequence converges. For the case where $0 < |r| < 1$, we need to show that there exists a positive number N such that for all $n > N$, $|r^n - 0| < \varepsilon$, no matter how small ε is. We can rewrite the inequality

$$|r^n - 0| < \varepsilon$$

as

$$|r|^n < \varepsilon.$$

Taking the logarithm of the inequality, we find that

$$n\ln|r| < \ln \varepsilon.$$

Recognizing that for $0 < |r| < 1$, $\ln |r|$ is negative, so that in dividing by $\ln |r|$, the inequality sign reverses direction, we find that

$$n > \frac{\ln \varepsilon}{\ln |r|}.$$

This inequality will help us investigate the condition on n ($n > N$) for which the sequence will converge. In the case where $\varepsilon \geq 1$, $(\ln \varepsilon / \ln |r|) \leq 0$. If we now let

$$N = \frac{\ln \varepsilon}{\ln |r|},$$

then the inequality will be true whenever $n > N$ (where N is any positive number). For the case where $\varepsilon < 1$, $\ln \varepsilon < 0$ and $(\ln \varepsilon / \ln |r|) > 0$. Therefore, whenever $n > N > 0$, the inequality is true, and the sequence converges. In short, we have used the definition of convergence to prove that

$$\lim_{n \to \infty} r^n = 0 \text{ for } |r| < 1.$$

As this example demonstrates, proving that a sequence converges based directly on the definition of convergence can become a bit involved, and this is where limit theorems, which we will discuss next, become useful.

1.4.1 Limit Theorems

Limit theorems are useful because, in practice, it can be difficult to establish the convergence of a sequence based solely on the definition of convergence. The use of limit theorems makes it possible to evaluate difficult limits by considering the limits of simpler or known sequences. In some cases, it is even possible to prove that a sequence is convergent without determining its limit.

Infinite sequences can be combined to form sum and difference sequences as well as product and quotient sequences. Adding or subtracting the corresponding terms of two sequences forms a sum or difference sequence. Similarly, multiplying or dividing the corresponding terms of two sequences (in the latter case providing no division by zero occurs) forms a product or quotient sequence.

Convergent sequences have the useful property that they can be combined to form other convergent sequences. The resulting sum, difference, product, or quotient sequence will have a limit that can be found through the use of limit theorems. The limit theorems that follow are given without proofs, which can be found in [2].

If $\{a_n\}$ and $\{b_n\}$ are convergent, infinite sequences with limits a and b, respectively, then

Theorem 1.1
$$\lim_{n\to\infty} (a_n + b_n) = a + b.$$

Theorem 1.2
$$\lim_{n\to\infty} (a_n - b_n) = a - b.$$

Theorem 1.3
$$\lim_{n\to\infty} (a_n b_n) = ab.$$

Theorem 1.4
$$\lim_{n\to\infty} c a_n = c \lim_{n\to\infty} a_n .$$

Theorem 1.5
$$\lim_{n\to\infty} a_n^c = \left[\lim_{n\to\infty} a_n\right]^c.$$

Theorem 1.6
$$\lim_{n\to\infty} (a_n/b_n) = \frac{a}{b} \ \ if \ b \neq 0 .$$

Theorem 1.7
$$\lim_{n\to\infty} r^n = 0 \ if\,|r| < 1.$$

Theorem 1.8
$$\lim_{n\to\infty} |r^n| = \infty \ if\,|r| > 1.$$

Theorem 1.9 *A sequence is unique.*

Theorem 1.10 *If $a_n \leq b_n$ for all $n > N$, then $\lim_{n\to\infty} a_n \leq \lim_{n\to\infty} b_n$.*

Theorem 1.11 *If $a_n \leq s_n \leq b_n$ for all $n > N$ and $\lim_{n\to\infty} a_n = \lim_{n\to\infty} b_n = L$, then $\lim_{n\to\infty} s_n = L$.*

(This is referred to as the "Squeeze Theorem.")

Theorem 1.12 *If a_n is equal to a constant c for all n, then the sequence is $c, c, c, \ldots, c, \ldots,$ and $\lim_{n\to\infty} a_n = \lim_{n\to\infty} c = c$.*

As an example of how limit theorems are used to determine the limit of a sequence, consider the geometric sequence $\{ar^n\}$. To determine the convergence properties of $\{ar^n\}$, we can attempt to construct a proof using the definition of convergence, as we did at the end of Section 1.4, or we can use limit theorems, as follows. By combining limit theorems 1.4 and 1.7, we can determine the following condition under which the limit of the general term ar^n of a geometric sequence is zero:

$$\lim_{n \to \infty} ar^n = 0 \text{ if } |r| < 1.$$

Therefore, a geometric sequence converges for $-1 < r < 1$. Furthermore, by combining limit theorems 1.4 and 1.8, we can conclude that a geometric sequence diverges when $|r| > 1$:

$$\lim_{n \to \infty} |ar^n| = \infty \text{ if } |r| > 1.$$

Thus, we have established the convergence and divergence conditions for geometric sequences by way of limit theorems. These properties of geometric sequences are frequently used in evaluating problems involving growth and decay.

1.4.2 Techniques for Dealing with Difficult Limits

Often, taking the limit of a general expression is straightforward. However, there are cases where the limit has one of the **indeterminate** forms $0/0$, ∞ / ∞, $0 \times \infty$, $\infty - \infty$, 0^0, 1^∞, or ∞^0. In these cases, further analysis is needed to determine whether or not a limit exists. In this section, we will explore techniques for dealing with these and other difficult-to-evaluate limits [3].

1.4.2.1 Handling Limits with the Indeterminate Form 0/0 or ∞ / ∞

L'Hôpital's Rule is a technique for handling limits containing a quotient $f(x)/g(x)$ that has the indeterminate form $0/0$ or ∞ / ∞. The technique consists of taking the derivatives of $f(x)$ and $g(x)$ separately and then evaluating the limit of the quotient $f(x)'/g(x)'$:

$$\lim_{n \to c} \frac{f(x)}{g(x)} = \lim_{n \to c} \frac{f(x)'}{g(x)'}.$$

As an example, consider the infinite sequence $\{\ln(n) / n\}$. To test this sequence for convergence, we take the limit of the nth term of the sequence as n gets large:

$$\lim_{n \to \infty} \frac{\ln n}{n}.$$

We discover that as $n \to \infty$,

$$\frac{\ln n}{n} \to \frac{\infty}{\infty} .$$

The quotient has the form ∞ / ∞, which is indeterminate. So further analysis is needed to determine whether a limit exists. Applying L'Hôpital's Rule to the quotient, we find that

$$\lim_{n\to\infty} \frac{\ln n}{n} = \lim_{n\to\infty} \frac{1/n}{1} = 0.$$

Therefore, the sequence is convergent and has the limit zero.

There are cases where L'Hôpital's Rule may need to be applied several times in a row. For example, in the limit

$$\lim_{n\to 0} \frac{e^n + e^{-n} - 2}{1 - \cos 2n} ,$$

the quotient has the indeterminate form $0 / 0$. Applying L'Hôpital's Rule once, we find that

$$\lim_{n\to 0} \frac{e^n - e^{-n}}{2 \sin 2n} .$$

Unfortunately, this quotient still has the indeterminate form $0 / 0$. Applying L'Hôpital's Rule a second time, we obtain

$$\lim_{n\to 0} \frac{e^n + e^{-n}}{4 \cos 2n} = \frac{1}{2} .$$

Therefore, the limit exists and is 1/2.

1.4.2.2 Handling Limits with the Indeterminate Form $0 \times \infty$ or $\infty - \infty$

If a limit has the indeterminate form $0 \times \infty$ or $\infty - \infty$, it may be possible to make an algebraic or some other type of change in the expression to bring it into the form $0/0$ or ∞ / ∞, whereupon L'Hôpital's Rule can be applied.

As an example, assume that taking the limit of the product of two functions leads to $0 \times \infty$:

$$\lim_{n\to c} f(n)\, g(n) \to 0 \times \infty .$$

By rewriting the product $f(n)g(n)$ as a ratio,

$$f(n)g(n) = \frac{g(n)}{\frac{1}{f(n)}} \text{ or } \frac{f(n)}{\frac{1}{g(n)}},$$

the indeterminate form

$$f(n)g(n) \to 0 \times \infty$$

can be rewritten as

$$\frac{g(n)}{\frac{1}{f(n)}} \to \frac{1}{\infty} \times \infty = \frac{\infty}{\infty} \ .$$

Alternatively, we may choose to rewrite

$$f(n)\ g(n) \to 0 \times \infty$$

as

$$\frac{f(n)}{\frac{1}{g(n)}} \to 0 \times \frac{1}{0} = \frac{0}{0} \ .$$

In either case, we are now able to apply L'Hôpital's Rule. As an example, in taking the limit of the sequence $\{n \sin (1/n)\}$, we would find that

$$\lim_{n \to \infty} n \ \sin (1/n) \to \infty \times 0 \ ,$$

which is indeterminate. It is possible to rewrite the expression so that it appears as the ratio

$$\lim_{n \to \infty} \frac{\sin (1/n)}{1/n} \to \frac{0}{0} \ .$$

It is now possible to apply L'Hôpital's Rule and find that

$$\lim_{n \to \infty} \frac{\sin(1/n)}{1/n} = \lim_{n \to \infty} \frac{(1/n^2)\cos(1/n)}{1/n^2} = \lim_{n \to \infty} \cos(1/n) = 1.$$

1.4.2.3 Handling Limits with the Indeterminate Form 0^0, 1^∞, or ∞^0

For expressions with the indeterminate form 0^0, 1^∞, or ∞^0, taking the expressions' natural logarithm may bring them to the form $0 \times \infty$ or $\infty \times 0$, which can be handled using the methods presented in previous sections.

The indeterminate forms 0^0, 1∞, and ∞^0 typically occur when taking the limit of expressions of the form

$$y \ (x) = f(x)^{g(x)}. \tag{1.6}$$

Taking the natural logarithm of both sides of the expression, we obtain

$$\ln y(x) = g(x)\ln f(x) \ . \tag{1.7}$$

If the function $y(x)$ has the indeterminate form 0^0 or ∞^0, then the function $\ln y(x)$ has the indeterminate form $0 \times \infty$. If $y(x)$ has the indeterminate form 1^∞, then $\ln y(x)$ has the form $\infty \times 0$. The limits of $0 \times \infty$ and $\infty \times 0$ can then be evaluated using the methods described in the previous sections.

It is important to recognize that evaluating $0 \times \infty$ or $\infty \times 0$ provides the limit of the function $\ln y(x)$, but we were initially seeking the limit of the function y (x). So how does knowing the limit of $\ln y(x)$ help us determine the limit of $y(x)$? If the function $\ln y(x)$ has the limit L, i.e.,

$$\lim_{x \to \infty} \ln y(x) = \lim_{x \to \infty} g(x) \ \ln f(x) = L \ ,$$

then the function $y(x)$ has the limit e^L:

$$\lim_{x \to \infty} y(x) = \lim_{x \to \infty} e^{\ln y(x)} = \lim_{x \to \infty} f(x)^{g(x)} = e^L.$$

As an example, consider testing the infinite sequence $\{(1+n^2)^{1/\ln n}\}$ for convergence. To test for convergence, we take the limit of the nth or general term of the sequence. In evaluating the limit, we discover that the general term has the indeterminate form ∞^0:

$$\lim_{n \to \infty} (1 + n^2)^{\frac{1}{\ln n}} \to \infty^0.$$

So further analysis is needed in order to determine whether the sequence approaches a limit. Recognizing the indeterminate form ∞^0, we use the natural logarithm of $(1+n^2)^{1/\ln n}$ to obtain:

$$\ln \left((1 + n^2)^{\frac{1}{\ln n}} \right) = \frac{1}{\ln n} \ln(1 + n^2).$$

In evaluating the limit of this expression, we find that it is 2:

$$\lim_{n \to \infty} \frac{1}{\ln n} \ln(1 + n^2) \to \frac{\ln(n^2)}{\ln n} \to \frac{2\ln n}{\ln n} \to 2.$$

Therefore, our original expression $(1+n^2)^{1/\ln n}$ has the limit e^2, i.e.,

$$\lim_{n \to \infty} (1 + n^2)^{\frac{1}{\ln n}} = \lim_{n \to \infty} e^{\left(\frac{1}{\ln n} \ln(1+n^2) \right)} = e^2 \ ,$$

and the sequence is thus convergent.

1.4.2.4 Handling Limits Containing the Factorial $n!$

Applying the previous methods to indeterminate forms involving a factorial $n!$ would require taking the derivative or finding the natural logarithm of the factorial. In such cases, it may be better to make an algebraic substitution so that the factorial appears in a more manageable or recognizable form. Consider the following limit:

$$\lim_{n \to \infty} \frac{10^n}{n!} \; .$$

While the quotient $10^n/n!$ has the indeterminate form ∞/∞, applying L'Hôpital's Rule would involve taking the derivative of a factorial. Instead, we can try expanding or reexpressing the quotient as

$$\frac{10^n}{n!} = \frac{10^{10}}{10!} \times \left(\frac{10}{11}\right) \times \left(\frac{10}{12}\right) \times \left(\frac{10}{13}\right) \ldots \times \left(\frac{10}{n}\right),$$

where the first term of the expression, $10^{10}/10!$, represents the first ten products and the remaining terms represent the last $n - 10$ products, or $(10/n)^{n-10}$. Incorporating these observations, we can rewrite the expression as

$$\frac{10^n}{n!} = \frac{10^{10}}{10!} \times \left(\frac{10}{n}\right)^{n-10} = \frac{10^{10}}{10!} \times \left(\frac{10}{11}\right) \times \left(\frac{10}{12}\right) \times \left(\frac{10}{13}\right) \ldots \times \left(\frac{10}{n}\right).$$

Consider the last $n - 10$ products of the expression, starting with the term $10/11$. Notice that $10/11$ is larger than any of the terms that follow it: $10/12, 10/13, \ldots,$ $10/n$. If we were to go term by term, replacing each of these lesser terms with $10/11$, the resulting product would clearly be larger:

$$\frac{10^{10}}{10!} \times \left(\frac{10}{11}\right) \times \left(\frac{10}{12}\right) \times \left(\frac{10}{13}\right) \ldots \times \left(\frac{10}{n}\right) < \frac{10^{10}}{10!} \times \left(\frac{10}{11}\right) \times \left(\frac{10}{11}\right)$$
$$\times \left(\frac{10}{11}\right) \ldots \times \left(\frac{10}{11}\right).$$

This is expressed more compactly as

$$\frac{10^n}{n!} = \frac{10^{10}}{10!} \times \left(\frac{10}{n}\right)^{n-10} < \left(\frac{10}{10!}\right) \times \left(\frac{10}{11}\right)^{n-10}.$$

We also need to recognize that the index n starts with $n = 1$, and therefore $10^n/n!$ is a positive number, i.e.,

$$0 < \frac{10^n}{n!} < \left(\frac{10^{10}}{10!}\right) \times \left(\frac{10}{11}\right)^{n-10}.$$

We are now in position to evaluate the limit of $10^n/n!$ by evaluating the limit of the inequality

$$\lim_{n\to\infty} 0 < \lim_{n\to\infty} \frac{10^n}{n!} < \lim_{n\to\infty} \left(\frac{10^{10}}{10!}\right) \times \left(\frac{10}{11}\right)^{n-10}.$$

We note that if the left- and right-hand sides of the inequality both approach the same limit, then, by the Squeeze Theorem (1.9), the middle expression also approaches that limit. The limit of the left-hand side of the inequality is 0, by eq. (1.10), as is the right-hand side of the inequality. Since $10/11 < 1$,

$$\lim_{n\to\infty} 0 \to 0 \quad \text{and} \quad \lim_{n\to\infty} \left(\frac{10^{10}}{10!}\right) \times \left(\frac{10}{11}\right)^{n-10} \to \frac{10^{10}}{10!} \times 0 = 0.$$

Since both limits equal zero, by the Squeeze Theorem we have

$$\lim_{n\to\infty} \frac{10^n}{n!} \to 0.$$

1.5 Examples

Chapter 1 will close with some examples showing how sequences occur in a variety of different contexts.

1.5.1 The Mariner's Rule of Twelfths

The Mariner's Rule of Twelfths is a sequence that serves as a mnemonic device. It is used to estimate the depth of shallow water at any time between a known low and high tide [4]. Expressed as a sequence, the Rule of Twelfths is 1, 2, 3, 3, 2, 1.

The cyclical period of ocean tides is dictated by the Earth's rotation. Semidiurnal low and high tides are approximately six hours apart. Tide tables give the time and expected height of both the low and high tides. However, being able to estimate the height of the tide at a time that falls between the low and high tide is important for navigating ships in shallow water. This is where the Rule of Twelfths is useful.

The rate at which tides rise and fall is not constant. Tide heights change slowly at the beginning and end of a tidal cycle and rapidly in the middle of the cycle. During the first hour after a low or high tide, the water level will change (rise or fall) by 1/12 of the total height of the tide. During the second hour, the water level will change by another 2/12 of the tidal height. The water level changes by 3/12 during the third hour and 3/12 in the fourth hour. The change in tide height then slows to 2/12 in the fifth hour and 1/12 in the sixth hour. This hourly change in height forms the sequence 1/12, 2/12, 3/12, 3/12, 2/12, 1/12, which is traditionally expressed in twelfths as 1, 2, 3, 3, 2, 1. Each term of the sequence represents the change in the water level during a one-hour period of a six-hour tidal cycle.

For example, if a tide table displays the heights of consecutive low and high tides as −0.1 m and +1.9 m, respectively, then the total height or range of the tide will be the difference between these two heights, i.e., 2 m. So in the first hour after a low or high tide, the water level will change by 1/12 × 2 m. During the second hour, the water level will change by an additional 2/12 × 2 m, during the third hour by an additional 3/12 × 2 m, and so on.

1.5.2 Photographic F-Stops

The sequence of f-stops 1, 1.4, 2, 2.8, 4, 5.6, 8, 11, 16 found on the aperture ring of a camera lens is approximately a geometric sequence. The common ratio between consecutive f-stops is approximately $\sqrt{2}$, with the value of the square root rounded off to make the ratios easier to print and remember.

An f-stop represents the ratio between the sizes of the various lens apertures. Closing the aperture or lowering by one f-stop, say, going from f/16 to f/22, cuts the amount of light entering the camera in half, while opening the aperture or increasing by one f-stop, say, going from f/8 to f/5.6, doubles the amount of light entering the camera.

To better understand how the factors 2, 1/2, and $\sqrt{2}$ are related, consider the area A_1 of a circular aperture, which is given by

$$A_1 = \pi r_1^2.$$

The area for a circular aperture A_2 with twice the area of A_1 is given by

$$A_2 = \pi r_2^2 = 2A_1 = 2\pi r_1^2$$

and the ratio between areas A_2 and A_1 is given by

$$A_2/A_1 = 2 = \frac{\pi r_2^2}{\pi r_1^2},$$

while the relationship between the radii r_1 and r_2 is

$$r_1\sqrt{2} = r_2.$$

To double the area of a circular aperture, the radius of the aperture needs to be increased by a factor of $\sqrt{2}$. To halve the area of a circular aperture, the radius needs to be reduced by a factor of $1/\sqrt{2}$. It is worth noting that the common shutter speeds on a camera (in seconds) are 1, 1/2, 1/4, 1/8, ..., which is also a geometric sequence, and that indexing up or down by one setting either doubles or halves the exposure time.

1.5.3 Wires in Conduit

Electrical conduit is tubing that is used to enclose and protect wires. When the interior diameter D_n of conduit is indexed with n, the maximum number of identical wires that can be enclosed by conduit diameters D_n form a sequence [5]. If the enclosed wires each have the diameter d, then the sequence $\{D_n\}$ of conduit diameters is given by

$$D_1, D_2, D_3, D_4, D_5, \ldots = d, 2d, \left(1 + \frac{2}{\sqrt{3}}\right)d, \left(1 + \sqrt{2}\right)d,$$

$$\frac{1}{4}\sqrt{2(5 - \sqrt{5})}d, \ldots = \{D_n\}\ .$$

Figure 1.2 illustrates how wires are configured inside conduit.

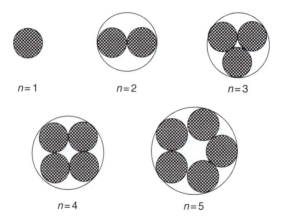

$n=1$ $n=2$ $n=3$

$n=4$ $n=5$

Figure 1.2 Identical wires enclosed in conduit of diameter D_n.

1.5.4 Life Predates the Formation of Earth

It has been suggested that the genetic complexity of organisms doubles every 376 million years [6]. If this rate is assumed to be constant, then genetic complexity would conform to a geometric sequence with a period of 376 million years and a common ratio $r = 2$.

If g_t represents the genetic complexity at time t and g_{t+1} represents the genetic complexity one doubling period later, then

$$\frac{g_{t+1}}{g_t} = 2,$$

and genetic diversity can be written as the geometric sequence:

$$g_t = g_0(2)^t.$$

If the genome size (the number of nucleotide base pairs) is taken to be the measure of genetic diversity and one assumes an initial genetic diversity g_0 of one base pair, then the law of geometric growth can be used to extrapolate backward in time to estimate when life began to evolve. Taking the logarithm of our expression for genetic diversity, we find that

$$\ln (g_t) = t \ln (2).$$

Solving for t, the number of doubling periods that have transpired, in order to reach the current level of genetic diversity, we find that

$$t = \frac{\ln (g_t)}{\ln (2)}.$$

Estimating the current genome size to be on the order of 10^8 base pairs, the number of doubling periods that have transpired since life began is

$$\frac{\ln (10^8)}{\ln (2)} \simeq \frac{18}{0.69} \simeq 26.$$

With a doubling period of 376 million years, this implies that life began evolving approximately 26×375 million, or 9.7 billion, years ago. Earth is currently believed to be 4.5 billion years old, which suggests that life predates the formation of Earth by about 5.2 billion years. While this calculation is simply a back-of-the envelope estimation and is by no means conclusive, it is intriguing to consider the possibility that life began evolving in the universe even before Earth was formed.

1.5.5 Chain Reactions: Avalanches, Viruses,
and Geometric Growth

Geometric growth occurs frequently in many real-world situations, so it is important to develop a sense of how such growth progresses and the impact that the growth factor, r, has on the size of the nth generation. We will consider a few examples where we can easily compare the impact of the growth factor r.

A chain reaction is a self-sustaining sequence of events. For example, consider a single falling rock dislodging two rocks, which each dislodge another two rocks, and so on. . . . This would result in runaway geometric growth and the beginning of an avalanche.

Chain reactions can be modeled as geometric sequences in which the size of a generation a_{n+1} is proportional to the size of the previous generation a_n:

$$a_{n+1} = a_n r \quad \text{or} \quad a_n = ar^{n-1},$$

where r (the common ratio) is referred to as the growth factor. We know that for $r > 1$, a geometric progression will increase without a bound (will diverge), while for $0 < r < 1$, a geometric progression will converge to zero, i.e., the reaction will come to a stop.

A very dramatic example of a chain reaction is a nuclear explosion. When a neutron hits a uranium-235 nucleus, there is a high probability that the nucleus will split (fission) into two smaller nuclei, while releasing energy and two additional neutrons. Both neutrons are capable of inducing fission reactions in other uranium nuclei, potentially starting a chain reaction. Neutron-induced fission in uranium-235 has a growth factor $r = 2$. The first-generation fission can induce two second-generation fissions, which in turn can induce four third-generation fissions, which in turn can induce eight fourth-generation fissions, and so forth. This is clearly the geometric progression

$$1, \ 2, \ 4, \ 8, \ldots, \ 2^{n-1}, \ldots$$

Neutron-induced fission occurs in other elements as well, but with different growth factors. For example, the neutron-induced fission of plutonium-239 releases three neutrons. In this case the growth factor is $r = 3$, and the geometric progression for plutonium-239 looks like:

$$1, \ 3, \ 9, \ 27, \ldots, \ 3^{n-1}, \ldots$$

How long would it take the plutonium-239 chain reaction to consume 1 kg of plutonium fuel? How long would it take the uranium-235 chain reaction to consume 1 kg of uranium fuel? There are approximately 2.6×10^{24} nuclei in 1 kg of both plutonium-239 and uranium-235. For uranium, the growth factor is

$r = 2$, and therefore all the nuclei in 1 kg would be consumed by the chain reaction in about $n = 81$ generations:

$$2.6 \times 10^{24} \cong 2^{81}.$$

Fission occurs in approximately 10^{-8} s, so it would take about 0.8×10^{-6} s to completely fission 1 kg of uranium-235. In the case of plutonium-239, where $r = 3$, it would take about 51 generations:

$$2.6 \times 10^{24} \cong 3^{51},$$

or about 0.5×10^{-6} s, for the chain reaction to consume 1 kg of plutonium-239.

Now consider a software virus that is capable of spreading to 10 unprotected devices via the Internet, each of which is able to spread to another 10 devices. Virus populations follow geometric growth, and in this case, the growth factor is $r = 10$, so that

$$a_n = 10^{n-1}.$$

How many generations would it take for such a virus to contact every device on the Internet? If we assume that the number of devices connected to the "Internet of things" is approximately 10×10^9, it would take about $n = 9$ generations for the virus to contact every device.

It is important to recognize the impact of the growth ratio r on the avalanching or "snowballing" property of geometric growth. In each of these examples, the percentage change for the total population during the final generation is given by

$$\left(1 - \frac{1}{r}\right) \times 100\%.$$

For $r = 2$ in the case of nuclear fission, half of the total amount of fuel is consumed during the final generation. The other half was consumed by the combined previous generations. For $r = 3$, approximately 66% of the total amount of fuel is consumed during the final generation, and the previous generations combined consumed only about 33% of the total. For $r = 10$ in the case of viruses, approximately 90% of all of the infections occur during the final generation, while all previous generations combined were responsible for about only 10% of the total number of infections. This example illustrates the impact of the growth factor r and what is meant by the expression "going viral."

A simpler illustration of a "snowballing effect" can be found in the hypothetical water lily problem [8]. Consider a pond containing invasive water lilies. If the water lilies are allowed to grow unchecked, they will eventually cover the pond and kill all the other life in the pond. Consider a water lily population that doubles every day (a growth factor of $r = 2$). If the lilies are allowed to grow until they cover half the pond, we will only have one day to save the pond. If the water lilies have a growth factor $r = 3$, then by the time the lilies cover one-third of the pond, we will only have one day to respond and save the pond. Finally, with a growth factor of $r = 10$, by the time the water lilies cover one-tenth of the pond, we will only have a single day to respond.

Before we end this chapter, we should clarify the subtle difference between geometric and exponential growth. In both cases, the rate of growth (or decay) is proportional to the current value or size of a population. However, **geometric growth** describes cases where incremental changes in a population take place at discrete and equal time intervals. Examples of geometric growth include animal populations where the species has a breeding season and the compounding of interest, which happens periodically, such as once a month or once a quarter (in three-month intervals). **Exponential growth** refers to cases where a population grows continuously. Bacteria and human populations provide examples of exponential growth. Neither humans nor bacteria have discrete breeding seasons, and therefore both reproduce more or less continuously.

2

Infinite Series

2.1 Introduction to Series

Infinite series expressions are frequently used in mathematics, engineering, and the sciences, particularly in problems where an exact or closed-form solution cannot be found. Infinite series methods are used to solve or approximate the solutions to nonelementary integrals, differential equations, and partial differential equations. Many elementary functions can be represented as an infinite series. For example, the value of a transcendental function can be found by evaluating its infinite series representation.

An **infinite sequence** is defined as an infinite succession of terms $a_1, a_2, a_3, \ldots$ An **infinite series** is the summation of an infinite number of terms and is represented by

$$a_1 + a_2 + a_3 + \cdots + a_n + \cdots \tag{2.1}$$

In cases where we are able to identify the rule or algorithm that produces the series or express the nth or general term of the series using a formula, it is possible to represent the series using **sigma** or **summation notation:**

$$\sum_{n=1}^{\infty} a_n \text{ or } \sum a_n.$$

For expressions where the limits are omitted, as in the right-hand expression, it is understood that the summation is over the range from $n = 1$ to $n = \infty$. The reader should be aware that indexing does not necessarily have to begin with the number 1. Sometimes it is useful to change or shift indexing in order to simplify an expression or to combine multiple series expressions under a single sigma sign. For example, consider summing the following series:

$$\sum_{k=2}^{\infty} b_{k-1}x^{k-1} + \sum_{n=1}^{\infty} a_n x^n.$$

By letting $n = k - 1$ in the second series, we can shift the indices so that the exponent of x is $k - 1$ in both series:

$$\sum_{k=2}^{\infty} b_{k-1}x^{k-1} + \sum_{n=1}^{\infty} a_n x^n = \sum_{k=2}^{\infty} b_{k-1}x^{k-1} + \sum_{k-1=1}^{\infty} a_{k-1}x^{k-1}$$

$$= \sum_{k=2}^{\infty} (b_{k-1} + a_{k-1})x^{k-1}.$$

Index shifting is frequently used when seeking a series solution to a differential equation; this topic will be covered in Chapter 5.

A **geometric series** is constructed from the terms of a geometric sequence and is written as

$$a + ar + ar^2 + ar^3 + \cdots + ar^n + \cdots = \sum_{n=1}^{\infty} a_n r^{n-1}. \tag{2.2}$$

Many infinite series do not have a sum. That is, the infinite series or infinite summation does not converge to a finite numerical value. If an infinite series has a definite value, it is called **convergent**; otherwise, it is called **divergent**. In applied problems, we are generally seeking a numerical value, for example: How long will it take for the value of an investment to double? How many times do I need to wash a container in order to remove 99% of the unwanted solution? We are interested in solving problems using methods that will converge to a solution or an approximation to the solution. For this reason, this text will focus almost exclusively on **convergent series** and leave **divergent series** to mathematicians. One exception will be **semiconvergent** or **asymptotic series**, which have applications in physics and computing.

2.2 Convergence and the Sequence of Partial Sums

While the concept of an infinite summation appears straightforward, there is a subtlety in that an infinite summation cannot be evaluated by simply adding consecutive terms. To better understand this point, consider a finite summation. A finite summation is calculated by simply adding consecutive terms. If there are too many terms to be summed by hand, we can use a computer to complete the summation. The difficulty with an infinite summation is that no matter how many thousands of terms are summed, infinitely many terms remain to be added. So a direct algebraic approach cannot work, even if it is executed by a computer.

Some method other than consecutive addition must therefore be used to determine the sum of an infinite series (if it has one). One such method is the **method of partial sums**. A sequence of partial sums $\{S_n\}$ is constructed from the terms of the infinite summation. The techniques developed in Chapter 1 can then be used to determine whether the sequence of partial sums, $\{S_n\}$, approaches a limit. If so, then the infinite series converges to a sum, and this is equal to the limit of the sequence of partial sums.

Consider the infinite series represented by

$$a_1 + a_2 + a_3 + \cdots + a_n + \cdots$$

We can determine whether this series has a sum by evaluating the limit of its sequence of partial sums, $\{S_n\}$. The elements of the sequence of partial sums are found by consecutively adding successive terms of the series:

$$
\begin{aligned}
S_1 &= a_1 \\
S_2 &= a_1 + a_2 \\
S_3 &= a_1 + a_2 + a_3 \\
S_4 &= a_1 + a_2 + a_3 + a_4 \\
&\cdots \\
S_n &= a_1 + a_2 + a_3 + a_4 + \cdots + a_n.
\end{aligned}
$$

S_1 is referred to as the first partial sum, S_2 the second partial sum, S_3 the third partial sum, and S_n the nth partial sum. These partial sums form an ordered list or sequence called the sequence of partial sums $\{S_n\}$, given by

$$S_1, S_2, S_3, \ldots, S_n, \ldots = \{S_n\}.$$

Recall from Chapter 1 that a sequence is convergent if the nth or general term approaches a finite limit, here denoted as S:

$$\lim_{n \to \infty} S_n = S.$$

How is the limit of the partial sums related to the sum of the infinite series? If the partial sums S_n approach a finite number S, then S is the **sum of the infinite series** and the series **converges**:

$$S = \lim_{n \to \infty} S_n = \lim_{n \to \infty} (a_1 + a_2 + a_3 + \cdots + a_n) \to \sum_{n=1}^{\infty} a_n. \tag{2.3}$$

Informally, a sequence of partial sums is convergent when the individual partial sums cluster together (approaching a limiting value) as you move out along the

Figure 2.1 Partial sums approaching a limit.

sequence. The limiting value of the partial sums is equal to the sum of the infinite series. Figure 2.1 provides a visual representation of how partial sums might converge to a limit.

It is important to note that the only way it is possible for individual partial sums S_n to approach a finite value is if the consecutive terms a_n of the infinite series approach zero, i.e.,

$$\lim_{n \to \infty} a_n = 0.$$

If this were not the case, the individual partial sums would not approach a limiting value. In order for an infinite series to be convergent, the terms a_n must approach 0 in the limit as n becomes large. This is a necessary but not a sufficient condition to ensure the convergence of a series. Put another way, convergent series all share a common feature, which is that the consecutive terms a_n of the series approach 0 as n becomes large, but unfortunately, some divergent series also have this property, so it does not guarantee convergence. We will revisit this condition in Section 2.3 when we consider convergence tests.

All repeating decimals, such as $0.333\ldots$ or $0.1851851\ldots$, are infinite series that illustrate convergence and the concept of the sequence of partial sums. Consider, for example, the fraction $1/3$. In decimal format, this fraction is expressed as $0.3333\ldots$, which can in turn be written as the infinite series

$$0.333\cdots = 0.3 + 0.03 + 0.003 + \cdots = \frac{3}{10} + \frac{3}{100} + \frac{3}{1000} + \cdots$$
$$+ \frac{3}{10^n} + \cdots = \sum_{n=1}^{\infty} \frac{3}{10^n}.$$

This infinite series demonstrates the property we have discussed; namely, in the limit as n approaches infinity, the terms of the infinite series approach zero:

$$\lim_{n \to \infty} \frac{3}{10^n} \to 0,$$

while the sequence of partial sums

$$\{S_1, S_2, S_3, S_4, \ldots\} = \{0.3, 0.33, 0.333, 0.3333, \ldots\}$$

approaches the limit 1/3:

$$\lim_{n \to \infty} S_n \to 0.3333\ldots = \frac{1}{3}.$$

2.2.1 Divergence

In cases where a sequence of partial sums S_n of an infinite series does not approach a limit but rather increases without bound, or oscillates, the infinite series is **divergent**. This text will not address divergent series except for the special cases of semiconvergent or asymptotic series. Asymptotic series are encountered as solutions to differential equations, in the evaluation of integrals, and in numerical computations. We will consider asymptotic series in Section 3.7.

2.2.2 Examples of Sequences of Partial Sums

In this section, the method of partial sums will be used to derive the convergence properties of **telescoping series** and **geometric series**.

Consider the infinite series

$$\sum_{n=1}^{\infty} \frac{1}{n(n+1)} = \sum_{n=1}^{\infty} \left(\frac{1}{n} - \frac{1}{(n+1)} \right) = \left(1 - \frac{1}{2} \right) + \left(\frac{1}{2} - \frac{1}{3} \right) + \cdots$$
$$+ \left(\frac{1}{n} - \frac{1}{(n+1)} \right) + \cdots.$$

In order to apply the method of partial sums, it is necessary to find a general expression for the sequence of partial sums. This general expression is then evaluated in the limit as $n \to \infty$ Writing out consecutive partial sums for the given series, we find that

$$S_1 = \left(1 - \frac{1}{2} \right)$$
$$S_2 = \left(1 - \frac{1}{2} \right) + \left(\frac{1}{2} - \frac{1}{3} \right)$$
$$S_3 = \left(1 - \frac{1}{2} \right) + \left(\frac{1}{2} - \frac{1}{3} \right) + \left(\frac{1}{3} - \frac{1}{4} \right)$$
$$\ldots$$
$$S_n = \left(1 - \frac{1}{2} \right) + \left(\frac{1}{2} - \frac{1}{3} \right) + \cdots + \left(\frac{1}{n-1} - \frac{1}{n} \right) + \left(\frac{1}{n} - \frac{1}{n+1} \right).$$

Observe that in each of these expressions for S_n, all of the inner terms cancel, leaving only the outer left- and right-hand terms:

$$S_n = \left(1 - \frac{1}{n+1}\right).$$

Series having this property are called **telescoping series**. Evaluating the limit of S_n in this example, we find that

$$\lim_{n\to\infty}\left(1 - \frac{1}{n+1}\right) = 1.$$

Therefore, the infinite series is convergent and sums to a value of 1.

As a second example, we will use the method of partial sums to establish the convergence properties of a **geometric series**

$$\sum_{n=1}^{\infty} ar^{n-1}, \qquad\qquad (2.4)$$

where a and r are real numbers and $a \neq 0$. We will find the convergence properties for four different conditions, specifically, in the domains where $|r| < 1$, $r = 1$, $r = -1$, and $|r| > 1$.

The expression for the nth partial sum of a geometric series is

$$S_n = a + ar + ar^2 + \cdots + ar^{n-1}.$$

We want to determine whether this partial sum approaches a limit. Unfortunately, this approach does not result in a general expression for S_n that can be evaluated in the limit for large values of n. However, there is a trick: if we multiply S_n by r, we get the expression

$$rS_n = ar + ar^2 + ar^3 + \cdots + ar^{n-1} + ar^n.$$

Subtracting the expression for rS_n from S_n gives

$$S_n - rS_n = a - ar^n.$$

Solving for S_n, we find that

$$S_n = \frac{a - ar^n}{1 - r}. \qquad\qquad (2.5)$$

This is a general expression for the sum of the first n terms (S_n) of a geometric series, and it can be evaluated in the limit as n approaches ∞:

$$\lim_{n\to\infty} S_n = \lim_{n\to\infty} \frac{a - ar^n}{1 - r}.$$

Notice that for the case where $|r| < 1$, the factor r^n in the numerator will approach 0 in the limit as n approaches infinity:

$$\lim_{n \to \infty} ar^n = 0.$$

Therefore, the individual partial sums S_n will approach the limit

$$\lim_{n \to \infty} S_n = \lim_{n \to \infty} \frac{a - ar^n}{1 - r} = \frac{a - 0}{1 - r} = \frac{a}{1 - r}.$$

We can therefore conclude that a geometric series is convergent when $|r| < 1$, and the sum of the geometric series is equal to $a / (1 - r)$.

Now consider the case where $r = 1$. In this case, the partial sum S_n is given by

$$S_n = a + a + a + \cdots + a(1)^n = na.$$

In the limit as n approaches ∞, the partial sum na grows without bound, i.e., the limit of S_n does not exist. So by definition, a geometric series is divergent when $r = 1$.

For the case where $r = -1$, the limit of the partial sum does not exist because the value oscillates between the values 0 and a:

$$S_n = a - a + a - a + \cdots + a(-1)^n.$$

Therefore, by definition, a geometric series is divergent when $r = -1$.

The fourth and final case is where $|r| > 1$. The quickest way to see that a geometric series will diverge in this domain is to recognize that the condition necessary for convergence, i.e., that the limit of the consecutive terms of the series must approach 0, is not fulfilled:

$$\lim_{n \to \infty} ar^n \neq 0.$$

The limit of ar^n does not approach 0 when $|r| > 1$, so by definition the geometric series must diverge. It is interesting to note that the geometric series has two distinct behaviors in this domain. For the case where $r > 1$, the terms of the geometric series grow without bound, while when $r < -1$, the terms of the series become larger in magnitude but also alternate in sign, i.e., oscillate, and the series is therefore divergent.

In summary, we have used the method of partial sums to prove that the infinite geometric series

$$a + ar + ar^2 + \cdots + ar^{n-1} = \sum_{n=1}^{\infty} ar^{n-1}, \text{ with } a \neq 0,$$

is a convergent series that sums to

$$\frac{a}{1-r}$$

when $|r| < 1$, and it is a divergent series when $|r| \geq 1$. We have also determined that the sum (finite) of the first n terms of a geometric series is given by

$$S_n = \frac{a(1 - r^n)}{1 - r} \quad \text{when } |r| < 1$$

and

$$S_n = \frac{a(r^n - 1)}{r - 1} \quad \text{when } r > 1. \tag{2.6}$$

The preceding expressions are important as they appear frequently in growth and decay problems and in fields as diverse as biology and finance, such as when determining the size of a population after n generations or the value of an investment after n interest periods. Several examples at the end of this chapter will illustrate how these expressions occur in very different contexts.

2.3 Testing Infinite Series for Convergence

It is generally difficult to establish the convergence of an infinite series based solely on the definition of convergence. This is because, once again, it is not always easy or even possible to find a general expression for the partial sum S_n. In such cases, one of several methods other than the method of partial sums must be used.

These other methods make use of the nth term a_n of the series. These methods can establish whether a particular series is divergent or convergent but will not determine the sum of the series. Yes, that's right, in some cases it is possible to establish that an infinite series is convergent without being able to calculate its sum.

2.3.1 The Divergence Test

Probably the simplest test of whether a series is convergent or divergent is the test for divergence:

If $\lim_{n \to \infty} a_n \neq 0$, then the infinite series $\sum a_n$ is divergent [2].

We first encountered this condition in Section 2.2. Recall that if the nth term a_n of an infinite series does not approach 0 as $n \to \infty$, then the infinite summation cannot approach a finite limit, and the series therefore diverges. Warning: this test is a test for divergence and makes no statement regarding convergence. For example, if we find that the limit of a_n is equal to 0 as $n \to \infty$, then the test is indeterminate. In this case, we cannot say whether the series converges or diverges; rather, we must apply a different test.

Applying the divergence test to the p-series is an illuminating exercise, as the test will return both determinate and indeterminate results for different domain values of p. Consider the **p-series**

$$1 + \frac{1}{2^p} + \frac{1}{3^p} + \cdots + \frac{1}{n^p} + \cdots = \sum_{n=1}^{\infty} \frac{1}{n^p} \tag{2.7}$$

and the two associated theorems:

Theorem 2.1 *p-series converge for $p > 1$.*

Theorem 2.2 *p-series diverge for $p \leq 1$.*

The divergence test can be used to prove part of the second p-series theorem, namely, that the p-series is divergent for $p < 1$. For $p < 1$, the p-series takes the following form:

$$1 + \frac{1}{2^{-p}} + \frac{1}{3^{-p}} + \cdots + \frac{1}{n^{-p}} + \cdots = 1 + 2^p + 3^p + \cdots + n^p + \cdots = \sum_{n=1}^{\infty} n^p.$$

Applying the divergence test to the nth term of the series, we find that

$$\lim_{n \to \infty} n^p \neq 0.$$

The consecutive terms of the p-series do not approach zero; therefore, the series is divergent. The case where $p = 1$ is a special case in which the p-series takes the following form:

$$1 + \frac{1}{2} + \frac{1}{3} + \cdots + \frac{1}{n} + \cdots = \sum_{n=1}^{\infty} \frac{1}{n}.$$

This series is called the **harmonic series**. Applying the divergence test to the general term of the harmonic series, we find that

$$\lim_{n \to \infty} \frac{1}{n} = 0.$$

This result is indeterminate, so we cannot say whether the series is divergent or convergent; rather, we must find another test. In Section 2.3.3, we will use the integral test to show that the harmonic series is divergent.

2.3.2 Convergence Tests for Positive Term Series

We are going to temporarily limit ourselves to considering only **positive term series**, i.e., infinite series in which every term is positive ($a_n \geq 0$). We adopt this restriction in order to develop four of the most common convergence tests. The **positive term series tests** are tests that can be applied to the nth term of a positive series to determine whether the series is convergent or divergent. These tests are useful in that they can establish the convergence of a series without needing a formula for the nth partial sum S_n and even without being able to sum the series.

2.3.3 The Integral Test

This test can be used on positive term series when the consecutive terms of the series do not increase beyond some point in the series, i.e., when $0 < a_{n+1} \leq a_n$ for some $n > N$. If the nth term a_n of a series is expressed as a function $f(n)$, i.e.,

$$\sum_{n=1}^{\infty} a_n = a_1 + a_2 + a_3 + \cdots + a_n + \cdots = f(1) + f(2) + f(3) + \cdots$$
$$+ f(n) + \cdots = \sum_{n=1}^{\infty} f(n),$$

then when the function $f(n)$ is positive valued, continuous, and decreasing, the original series converges if the integral

$$\int^{\infty} f(n)dn \tag{2.8}$$

is finite, and it diverges if the integral is infinite. This test works because of the geometric relationship between an infinite summation and an improper integral. Both a summation and an integral can be thought of as representing areas. The integral test is a comparison test in which we compare the area of an infinite summation to the area under a curve, i.e., an integral. This test comes with a caveat: you must be able to solve the integral.

Consider applying the integral test to the harmonic series

$$1 + \frac{1}{2} + \frac{1}{3} + \cdots + \frac{1}{n} + \cdots = \sum_{n=1}^{\infty} \frac{1}{n}.$$

Note that this is a positive term series, and the consecutive terms of the series are not increasing. If we allow ourselves to view n as a continuous variable, rather than as a discrete index, then the nth term of the series can be thought of as a function of a continuous variable: $a_n = f(n) = 1/n$. Therefore, if the integral

$$\int^{\infty} f(n)\,dn = \int^{\infty} \frac{1}{n}\,dn$$

is infinite, the harmonic series diverges, and if the integral is finite, then the harmonic series converges. Evaluating the integral, we find that

$$\int^{\infty} \frac{1}{n}\,dn = \ln n \Big|^{\infty} = \infty;$$

i.e., the integral is infinite, and therefore the harmonic series diverges. Note that the integral is only evaluated at the upper limit. The lower limit of the integral can be taken at any point in the series or chosen so as to drop any finite number of terms. Neglecting a finite number of terms of an infinite series will not affect the convergence or divergence of the series. Neglecting the first three terms or the first three million terms may affect the sum of the series but not its convergence or divergence.

Recall (from Section 2.3.1) that the harmonic series is a special case of the p-series ($p = 1$). The divergence test established that a p-series diverges for $p < 1$ but produced an indeterminate result for $p = 1$. With the use of the integral test, we now know that the p-series diverges for $p \leq 1$.

In the following section, we will make use of the **comparison test** to prove that the p-series is convergent for $p > 1$, thereby proving the p-series theorems, which were given without proof in Section 2.3.1.

2.3.4 The Comparison Test

The **comparison test** establishes the convergence or divergence of an unfamiliar positive term series by comparing it to a known (convergent or divergent) positive term series. For example, let the series

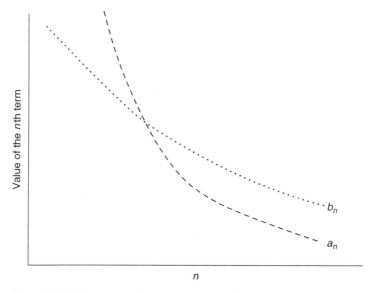

Figure 2.2 If $\sum b_n$ is known to be convergent, then $\sum a_n$ is convergent.

$$b_1 + b_2 + b_3 + \cdots = \sum_{n=1}^{\infty} b_n$$

be a convergent positive term series, and let the series

$$a_1 + a_2 + a_3 + \cdots = \sum_{n=1}^{\infty} a_n\mathrm{d}$$

be an unfamiliar series. If $a_n \le b_n$ for all integers n beyond some point, then the series $\sum a_n$ is convergent. Stated another way, a positive term series is convergent if it is **dominated** by a convergent series. The concept of **dominance** and how it relates to convergence are illustrated in Figure 2.2.

On the other hand, if the positive term series

$$b_1 + b_2 + b_3 + \cdots = \sum_{n=1}^{\infty} b_n$$

is known to be divergent and the terms of the unfamiliar series

$$a_1 + a_2 + a_3 + \cdots = \sum_{n=1}^{\infty} a_n$$

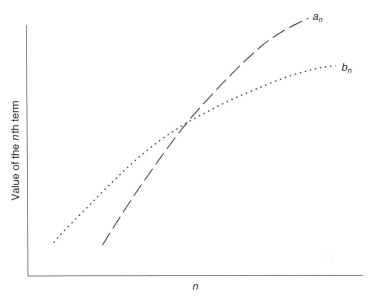

Figure 2.3 If $\sum b_n$ is known to be divergent, then $\sum a_n$ is divergent.

are such that $a_n \geq b_n$ for all integers n beyond some point, then the series $\sum a_n$ is divergent. Put another way, a positive term series is divergent if it **dominates** a divergent series. Figure 2.3 illustrates the concept of **dominance** and how it relates to divergence.

Be careful to not misapply the comparison test. If an unfamiliar series dominates a convergent series, the unfamiliar series may be convergent or divergent. Similarly, if an unfamiliar series is dominated by a divergent series, the unfamiliar series may be convergent or divergent. In both of these cases, the comparison test shows us nothing.

As an example, we will use the comparison test to prove that the p-series is convergent for $p > 1$. Writing out the expression for the p-series, we find that

$$\sum_{n=1}^{\infty} \frac{1}{n^p} = 1 + \frac{1}{2^p} + \frac{1}{3^p} + \cdots.$$

Observe that for $p > 0$,

$$\frac{1}{3^p} < \frac{1}{2^p},$$

and therefore

$$\left(\frac{1}{2^p} + \frac{1}{3^p}\right) < \left(\frac{1}{2^p} + \frac{1}{2^p}\right) = \frac{2}{2^p}.$$

Thus, the sum of the first two terms of the p-series is less than $2/2^p$. Applying the same logic to the next four terms of the p-series, we that find that

$$\left(\frac{1}{4^p} + \frac{1}{5^p} + \frac{1}{6^p} + \frac{1}{7^p}\right) < \left(\frac{1}{4^p} + \frac{1}{4^p} + \frac{1}{4^p} + \frac{1}{4^p}\right) = \frac{4}{4^p}.$$

Continuing this process with the next eight terms of the p-series produces

$$\left(\frac{1}{8^p} + \frac{1}{9^p} + \frac{1}{10^p} + \frac{1}{11^p} + \frac{1}{12^p} + \frac{1}{13^p} + \frac{1}{14^p} + \frac{1}{15^p}\right) < \frac{8}{8^p}.$$

Therefore, we conclude that

$$\sum_{n=1}^{\infty} \frac{1}{n^p} = 1 + \frac{1}{2^p} + \frac{1}{3^p} + \cdots < 1 + \frac{2}{2^p} + \frac{4}{4^p} + \frac{8}{8^p} + \cdots = \sum_{n=1}^{\infty} \left(\frac{2}{2^p}\right)^{n-1}.$$

Rewriting this expression gives us

$$\sum_{n=1}^{\infty} \frac{1}{n^p} < \sum_{n=1}^{\infty} \left(\frac{2}{2^p}\right)^{n-1}.$$

It is clear that our p-series is dominated by the geometric series when $a = 1$ and the common ratio is $r = 2/2^p$. If this geometric series is shown to be convergent, then the comparison test will lead us to conclude that the p-series is also convergent, since a series dominated by a convergent series is also convergent.

It is simple to demonstrate that the geometric series in this example converges. Recall that a geometric series is convergent when $|r| < 1$. In this example, $|r| = |2/2^p|$, which is less than 1 for $p > 1$. Therefore, the comparison test establishes that the p-series is convergent for $p > 1$.

The comparison test can be a very simple method for testing an unfamiliar series but only if we are able to identify a suitable comparison series. The harmonic series, the p-series, and the geometric series are now familiar series with known convergence properties. We can use these series to try to construct a comparison series suitable for testing an unfamiliar series. The harmonics series

$$\sum_{n=1}^{\infty} \frac{1}{n}$$

may be used to test an unfamiliar series for divergence, while the p-series may be used to test for either convergence or divergence because

$$\sum_{n=1}^{\infty} \frac{1}{n^p}$$

converges if $p > 1$ and diverges if $p \le 1$. The geometric series can also create a divergent or convergent comparison series;

$$\sum_{n=1}^{\infty} a(r)^{n-1}$$

converges if $|r| < 1$ and diverges if $|r| \le 1$.

2.3.5 The Ratio Test

Generally, it is difficult to apply the comparison or integral test to infinite series in which the general term contains the factorial $n!$ or the index n appears as an exponent, for example, $a_n = 3/n!$ or $a_n = 3^n$. On the other hand, the ratio of two consecutive terms containing either factorials and/or exponents is usually less complicated due to cancellation, for example:

$$\frac{a_{n+1}}{a_n} = \frac{\frac{3^{n+1}}{(n+1)!}}{\frac{3^n}{n!}} = \frac{3}{n+1}.$$

In such situations, the ratio test is easy to apply, so it is a frequently used test, even if it is not particularly sensitive.

Expressed in the form of a limit statement, the ratio test is as follows. Given

$$\rho = \lim_{n \to \infty} \left| \frac{a_{n+1}}{a_n} \right|, \tag{2.9}$$

if $\rho < 1$, the series converges, and if $\rho > 1$, the series diverges. If $\rho = 1$, the test is indeterminate, and a different test must be used.

The positive results (convergence or divergence) of the ratio test can be proven to be correct using the comparison test and the geometric series $1 + r + r^2 + r^3 + \dots$ [2]. We will not prove this here, but we can demonstrate the plausibility of the test by considering the consecutive terms of a geometric series. Recall that there is a common ratio r between consecutive terms of a geometric series:

$$\frac{a_{n+1}}{a_n} = r.$$

The geometric series converges if $|r| < 1$ and thus, in this case, when

$$\left|\frac{a_{n+1}}{a_n}\right| < 1,$$

i.e., when $\rho < 1$. On the other hand, the geometric series diverges if $|r| > 1$, thus when

$$\left|\frac{a_{n+1}}{a_n}\right| > 1,$$

i.e., when $\rho > 1$. This demonstrates the plausibility of the ratio test by considering the special case of a geometric series. The ratio test is, however, applicable to series in general, not just to geometric series.

There will be cases where the ratio test will return the indeterminate result $r = 1$. As an example, consider testing the harmonic series

$$1 + \frac{1}{2} + \frac{1}{3} + \cdots + \frac{1}{n} + \cdots$$

Taking the ratio of consecutive terms and evaluating the limit gives:

$$\rho = \lim_{n\to\infty} \left|\frac{\frac{1}{n+1}}{\frac{1}{n}}\right| = \lim_{n\to\infty} \frac{n}{n+1} = 1.$$

In this case, the ratio test is inconclusive, and another test must be used.

2.3.6 The Limit Comparison Test

The **limit comparison test** is the last positive term series test that we will consider. The limit comparison test states that if both $\sum a_n$ and $\sum b_n$ are positive term series and

$$\lim_{n\to\infty} \frac{a_n}{b_n} = c > 0,$$

then both series either converge or diverge. As with the ratio test, this test works because the ratio of two complicated expressions is frequently less complicated due to the cancellation of terms. The strategy for finding a suitable comparison series $\sum b_n$ is to take the expression for the unfamiliar series $\sum a_n$ and drop all but the largest factors contained in the numerator and the denominator of a_n. For example, consider

$$a_n = \frac{4n^2 + 6n}{3^n(n^2 + 5)}.$$

Neglecting all but the largest powers of n in both the numerator and the denominator, we obtain

$$a_n = \frac{4n^2}{3^n n^2} = \frac{4}{3^n}.$$

This would suggest taking $b_n = 1/3^n$ to be the nth term of the suitable comparison series for two reasons. First, choosing to consider only the largest factors in the expression for a_n guarantees that the a_n terms will be greater than the b_n terms, which is a condition for the test. Second, the b_n terms form a series with a known convergence property, namely, a convergent geometric series. Constructing the ratio of the two series and evaluating the limit, we find that

$$\lim_{n \to \infty} \frac{\frac{4n^2 + 6n}{3^n(n^2+5)}}{\frac{1}{3^n}} = \lim_{n \to \infty} \frac{3^n(4n^2+6n)}{3^n(n^2+5)} = \lim_{n \to \infty} \frac{4+6/n}{1+5/n^2} = 4 > 0.$$

Therefore, we can conclude that since $\sum b_n$ is convergent, $\sum a_n$ is also convergent.

2.3.7 Absolute Convergence

In Sections 2.3.3–2.3.6, we limited ourselves to considering only positive term series and the associated tests. The tests developed in those sections can be extended to analyze series that are not strictly positive. The convergence tests for positive terms series can be used to establish the property of absolute convergence for series containing negative terms.

If a series contains negative terms, for example,

$$\sum a_n = a_1 + a_2 - a_3 + a_4 + a_5 - a_6 + \cdots,$$

the related positive term series

$$\sum |a_n| = |a_1| + |a_2| + |a_3| + |a_4| + |a_5| + |a_6| + \cdots$$

is created by making each term of the original series positive (i.e., taking its absolute value). If the related positive term series can be shown to be convergent, then the original series containing negative terms is said to be **absolutely convergent**. It can be proven that a series that absolutely converges remains convergent when the negative signs are restored, although, of course, the series results in a different sum. To summarize: if an infinite series is absolutely convergent, then it is convergent.

For example, consider the **alternating infinite series**

$$0.3 - 0.03 + 0.003 - 0.0003\ldots = \frac{3}{10} - \frac{3}{100} + \frac{3}{1000} - \frac{3}{10000} + \cdots$$

$$= \sum_{n=1}^{\infty} \frac{(-1)^{n-1}3}{10^n}.$$

To test this series for absolute convergence, we first make all terms of the series positive, resulting in the series

$$0.3 + 0.03 + 0.003 + 0.0003 + \ldots = \frac{3}{10} + \frac{3}{100} + \frac{3}{1000} + \frac{3}{10000} + \cdots$$

$$= \sum_{n=1}^{\infty} \frac{3}{10^n}.$$

This series can then be tested for convergence. The series is recognizable as a geometric series that converges to 1/3. So we can conclude that the alternating series is absolutely convergent, and therefore convergent, although we would have to resort to some other method to sum the series.

To summarize in another way, if an infinite series of positive terms converges, then any related series in which some or all of the terms are negative will also be convergent. The sum of the series that includes negative terms will, of course, be less than the sum of the positive term series. As a simple example, consider the geometric series expression (a positive term series) for 1/3:

$$\sum_{n=1}^{\infty} \frac{3}{10^n} = 1/3.$$

If every other term of the series is made negative the series will still converge but to a smaller sum:

$$\sum_{n=1}^{\infty} \frac{(-1)^{n-1}3}{10^n} = 3/11.$$

Finally, if all the terms of the geometric series are made negative, the series will simply converge to the negative sum of the geometric series

$$\sum_{n=1}^{\infty} \frac{-3}{10^n} = -1/3.$$

Before we conclude the study of positive series, a few theorems about these series should be explicitly stated [8]:

Theorem 2.3 *If a positive term series converges, then the terms of the series may be rearranged in any order without affecting the sum or the convergence of the series.*

Theorem 2.4 *Convergent series may be added or subtracted term by term. The resulting series is convergent, and its sum is, respectively, the sum or the difference of the sums of the given series.*

Theorem 2.5 *The convergence or divergence of a series is unaffected by ignoring or changing a finite number of terms or multiplying every term in the series by a nonzero constant.*

2.4 Alternating Series

An infinite series whose terms alternate in their signs is called an **alternating series**, for example:

$$a_1 - a_2 + a_3 - a_4 + \cdots + (-1)^{n-1} a_n + \cdots \qquad (2.10)$$

If an alternating series converges, it converges rapidly due to the partial cancellation of consecutive terms of the series. This generally makes establishing convergence and approximating the sum for an alternating series simpler. It is also easy to determine the accuracy of an alternating series approximation/ representation of a solution.

The convergence test for alternating series is extremely simple. An alternating series converges if

$$|a_{n+1}| \leq |a_n| \text{ and } \lim_{n \to \infty} a_n = 0.$$

Informally speaking, an alternate series converges if the magnitude of the nth term approaches zero. For example, consider the **alternating harmonic series**:

$$1 - \frac{1}{2} + \frac{1}{3} - \frac{1}{4} + \cdots + \frac{(-1)^{n+1}}{n} + \cdots.$$

This series is convergent because

$$\left| \frac{1}{n+1} \right| \leq \left| \frac{1}{n} \right| \text{ and } \lim_{n \to \infty} \frac{1}{n} = 0.$$

Note that the alternating series test can only be applied to series in which the sign change is strictly alternating. Other tests, such as the integral test, should be applied when the sign change in the series is irregular or not strictly alternating.

The convergence properties of alternating series make them useful for approximating numerical values of functions and constructing numerical tables such as trigonometric tables. The nth partial sum S_n of an alternating series can be used as an approximation to the full sum S of the series, where the magnitude of the **error** of the approximation is less than the first neglected term a_{n+1} of the series:

$$|S - S_n| = |R_n| < a_{n+1}.$$

Here, R_n is referred to as the remainder of the series after n terms or the error in the approximation.

As an example, consider the series

$$1 - \frac{1}{2!} + \frac{1}{4!} - \frac{1}{6!} + \frac{1}{8!} - \cdots + \frac{(-1)^{n+1}}{n!} - \cdots$$

This series is convergent by the alternating series test: $a_{n+1} < a_n$ and $|a_n| = 1/n!$, as the limit is 0 as $n \to \infty$. If we approximate the full sum of the series using the fourth partial sum S_4, we find that

$$S \approx 1 - \frac{1}{2} + \frac{1}{24} - \frac{1}{720} \approx 0.5402.$$

The error in this approximation is less than the first neglected term, $a_5 = 1/8!$:

$$|R_n| < a_5 = \frac{1}{8!} \approx 0.00002,$$

and the approximation 0.5402 is accurate to about four decimal places. Note that the alternating series used in the preceding example is the series approximation for $\cos(1)$:

$$\cos(1) = 1 - \frac{1}{2!} + \frac{1}{4!} - \frac{1}{6!} + \frac{1}{8!} - \cdots + \frac{(-1)^{n+1}}{n!} - \cdots,$$

which demonstrates the use of an alternating series for approximating the value of a function.

2.5 Conditionally Convergent Series

Informally speaking, an infinite series can have exactly one of the following three properties: it can be **divergent**, it can be **absolutely convergent**, or it can be **conditionally convergent**. A divergent series is a series that does not approach a finite limit. An absolutely convergent series is a positive term series that converges. The convergence of an absolutely convergent series is not affected when terms are rearranged and/or their signs are altered. A **conditionally convergent** series is a series that converges but not absolutely. If the terms of a conditionally convergent series are rearranged or the signs are changed, the convergence and/or sum of the series may be affected. Such a series could even be made to diverge. Conditionally convergent series require careful treatment. When working with an unfamiliar convergent series, it should be treated as conditionally convergent unless and until it is proven to be absolutely convergent.

An infinite series

$$\sum a_n$$

is **conditionally convergent** if the series is convergent and

$$\sum |a_n|$$

is divergent.

The alternating harmonic series is an example of a conditionally convergent series. In Section 2.3.3, we used the integral test to establish the divergence of the harmonic series

$$1 + \frac{1}{2} + \frac{1}{3} + \cdots = \sum_{n=1}^{\infty} \frac{1}{n},$$

while in Section 2.4 we used the alternating series test to establish that the alternating harmonic series

$$1 - \frac{1}{2} + \frac{1}{3} - \cdots = \sum_{n=1}^{\infty} \frac{(-1)^{n+1}}{n}$$

is convergent. Therefore, the alternating harmonic series is not absolutely convergent but rather conditionally convergent. One way to better understand conditionally convergent series is to recognize that such a series is the sum of a divergent positive term series and a divergent negative term series. In taking the absolute values of the terms of a conditionally convergent series, we create a series that is the sum of two divergent series.

2.6 Examples

The following examples illustrate how the mathematical concepts of a series and convergence appear in different contexts.

2.6.1 Damped Oscillations

The motion of a damped oscillator can be described by an infinite geometric series. Consider releasing a rubber ball, pendulum, or mass on a spring from an initial displacement h. If the amplitude of each successive oscillation decreases, returning to only a fraction r of the previous amplitude, then for an infinite number of oscillations, the mass would travel the total distance T given by

$$T = h + 2hr + 2hr^2 + 2hr^3 + \cdots + 2hr^n + \cdots.$$

Figure 2.4 illustrates why there is factor of 2 for every term except the first.

Rewriting the expression for the total distance traveled using summation notation, i.e., as an infinite geometric series, we find that

$$T = h + \sum_{n=1}^{\infty} hr^{n-1}.$$

Summing the geometric series gives the result

$$T = h + \frac{h}{1 - r}$$

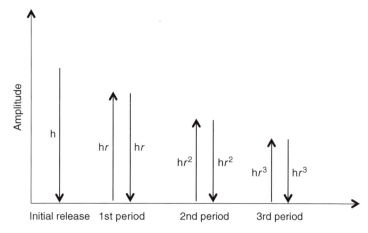

Figure 2.4 The periodic motion of a damped oscillator.

for $r < 1$. Allowing for an infinite number of oscillations is certainly physically unrealistic. To achieve something more realistic, one would truncate the series after a finite number of oscillations m, in which case the total distance traveled would be

$$T = h + \sum_{n=1}^{\infty} h r^{n-1} \cong h + \frac{h(1 - r^m)}{1 - r}.$$

To obtain this expression, the full sum of the infinite series was replaced with an approximation, namely, the expression for the mth partial sum.

2.6.2 Filtering

Consider a filtering process in which a fraction $1/m$ of an impurity is removed during each phase of the filtering [8]. That is, each successive phase removes $1/m$ of the amount removed by the preceding filter. Such a filtering process can be modeled using a geometric series. The reader may wish to revisit the example given in Section 1.3.3, where a geometric sequence was used to model the removal of an unwanted solution from a container.

If i is the initial amount of the impurity, then the total amount of impurity remaining after N filtering stages is given by

$$i - \frac{i}{m} - \frac{i}{m^2} - \frac{i}{m^3} - \cdots - \frac{i}{m^n} = i - \sum_{n=1}^{N} \frac{i}{m} \left(\frac{1}{m}\right)^{n-1}.$$

The series in our expression is a finite geometric series with a common ratio $r = 1/m$. Such a filtering process has an interesting feature. For $m = 2$, it is possible to remove as much of the impurity as we would like. However, for $m > 2$, it is not possible to remove all of the impurity, even if we filter indefinitely. To see this, consider running the filtering process indefinitely. In this case the infinite geometric series, with $|1/m| < 1$, would converge to the sum

$$S = \frac{i/m}{(1 - 1/m)} = \frac{i}{m - 1}.$$

Thus T, the total amount of impurity remaining after filtering indefinitely, is given by

$$T = i - \frac{i}{m - 1}.$$

For the case where $m = 2$,

$$T = i - \frac{i}{2-1} = i - i = 0,$$

the amount of impurity remaining approaches zero. That is, the impurity can be reduced to an arbitrarily low level through repeated filtering. On the other hand, for $m = 3$,

$$T = i - \frac{i}{3-1} = i - i/2 = i/2,$$

so the total proportion of impurity remaining would approach one-half of the original amount. Therefore, for a filtering process in which $m > 2$, it is not possible to reduce the impurity down to an arbitrarily low level, even if we filter indefinitely. Keep this example in mind the next time you proofread a paper: in order to approach an error-free paper, you need to be able to catch at least half the remaining errors during each reading.

2.6.3 Optical Cavities

In an optical cavity (such as a laser cavity), it is necessary for the light be reflected tens to hundreds of times between two end mirrors. Even a small decrease in a mirror's reflectivity percentage can greatly affect the performance of such a cavity. This can be demonstrated using an infinite series expression.

Assume that a large number of photons are bouncing back and forth between mirrors located at $x = 0$ and $x = 1$. If each mirror reflects a fraction r ($0 < r < 1$) of the incident photons, then $(1 - r)$ is the fraction of photons absorbed at each mirror. Starting with all photons at position $x = 0$ and traveling toward $x = 1$, the infinite series expression for the fraction of photons absorbed by the mirror at location $x = 1$ is given by

$$(1 - r) + r^2(1 - r) + r^4(1 - r) + \cdots = \sum_{n=1}^{\infty}(1 - r)(r^2)^{n-1} = \frac{1 - r}{1 - r^2}$$

$$= \frac{1 - r}{(1 - r)(1 + r)} = \frac{1}{1 + r},$$

while the fraction of photons absorbed by the mirror at $x = 0$ is

$$r(1 - r) + r^3(1 - r) + r^5(1 - r) + \cdots = \sum_{n=1}^{\infty}r(1 - r)(r^2)^{n-1} = \frac{r(1 - r)}{1 - r^2}$$

$$= \frac{r(1 - r)}{(1 - r)(1 + r)} = \frac{r}{1 + r}.$$

Therefore, the total number of photons absorbed by the two mirrors is

$$\frac{1}{1+r} + \frac{r}{1+r} = \frac{1+r}{1+r} = 1,$$

and so all of the photons will eventually be absorbed. The percentage of photons left in the cavity after n reflections is given by

$$(r)^{n-1} \times 100\%.$$

If the cavity mirrors were each 90% reflective ($r = 0.9$), then only about 50% of the photons would be left after 7 reflections ($n = 7$):

$$(0.9)^{7-1} \times 100\% \cong 53\%.$$

Therefore, 90% reflective mirrors will not allow for the tens to hundreds of reflections that may be needed to make the cavity function efficiently. If we need to have 50% of the photons remain in the cavity after 10 reflections, then the mirror reflectivity will need to be on the order of 93% ($r = 0.93$). Similarly, if the photons need to complete 100 reflections, then the mirror reflectivity will need to be 99%. As we can see, a small change in mirror reflectivity greatly affects the number of photons that will remain in the cavity.

3

Power Series

3.1 Interval of Convergence

Up to this point, we have only considered infinite series

$$S(n) = a_1 + a_2 + a_3 + \cdots + a_n + \cdots$$

in which each term is a constant. However, it is also possible to construct an infinite series in which the terms are functions of a variable rather than constants:

$$S(x) = f_1(x) + f_2(x) + f_3(x) + \cdots + f_n(x) + \cdots$$

There are a number of such infinite series that are important in engineering and the natural sciences. Examples of such series include power series, Fourier series, and Legendre series. Power series are expansions of the form,

$$f(x) = a_0 + a_1 x + a_2 x^2 + \cdots = \sum_{n=0}^{\infty} a_n x^n \qquad (3.1)$$

this type of series expansion is common as many mathematical functions have a **power series** representation.

The **Fourier series**,

$$S(x) = \frac{a_0}{2} + \sum_{n=1}^{\infty} a_n \cos nx + \sum_{n=1}^{\infty} b_n \sin nx \qquad (3.2)$$

is useful for representing periodic functions and forms the basis for harmonic analysis, which is important in the fields of communications, optics, and electrical engineering. The **Legendre series**,

$$S(x) = \sum_{n=0}^{\infty} t^n P_n(x) \qquad (3.3)$$

is a series of polynomials $P_n(x)$ called the **Legendre polynomials**; these are solutions to the Legendre differential equation, which occurs frequently in applied problems involving spherical symmetry.

The Legendre series expansion is also interesting in that it gives the best least-squares approximation to a polynomial fit. For example, if we are using a computer to fit a set of data points to a polynomial of degree n, i.e.,

$$f(x) = c_0 + c_1 x + c_2 x^2 + c_3 x^3 + \cdots + c_n x^n,$$

the expansion or representation of $f(x)$ as a series of Legendre polynomials (to order n) will always give the best least-squares fit.

Whether an infinite series of functions converges or diverges depends on the value of the variable x. In general, a series will converge for certain values of x and diverge for others. The values of x for which the series converges are called the **interval of convergence**. We have already encountered an example of this when considering the geometric series

$$S(r) = \sum_{n=1}^{\infty} ar^{n-1}.$$

Recall that the geometric series converges for $|r| < 1$ and diverges everywhere else, so that the interval of convergence for the geometric series is $-1 < r < 1$, with the common ratio r taken as the variable. In general the interval of convergence for a series will be expressed as $-r < x < r$, where r is referred to as the **radius of convergence.**

The concept of an interval of convergence is important, as it leads us to the idea of representing a function as an infinite series. To better understand the connection between convergent series and functions, it is important to recognize that a convergent series has a finite sum $S(x)$ for each value of x within the interval of convergence:

$$\sum_{n=1}^{\infty} f(x) = S(x), \text{ for each value of } x \text{ within the interval of convergence.}$$

Thus, for each point within the interval of convergence, the sum of the series can be expressed as a function $S(x)$. When the function $S(x)$ represents the sum of the infinite series, we have

$$\sum_{n=1}^{\infty} f(x) = S(x).$$

At this point, we are justified in adopting the point of view that the series converges to the function $S(x)$ and that the function $S(x)$ is represented by the series.

3.2 Properties of Power Series

Many mathematical functions have a **power series** expression. These types of infinite series occur frequently in engineering, mathematics, and physics and have the form

$$f(x) = a_0 + a_1 x + a_2 x^2 + \cdots = \sum_{n=0}^{\infty} a_n x^n \qquad (3.3)$$

when the function $f(x)$ is expanded around the origin or the form

$$f(x) = a_0 + a_1(x - c) + a_2(x - c)^2 + \cdots = \sum_{n=0}^{\infty} a_n (x - c)^n \qquad (3.4)$$

when the function is expanded around the point $x = c$.

The **interval of convergence for a power series** is typically found by the ratio test. Applying the ratio test to the power series expression

$$f(x) = \sum_{n=0}^{\infty} a_n x^n,$$

we find that the limiting ratio ρ is given by

$$\lim_{n \to \infty} \left| \frac{a_{n+1} x^{n+1}}{a_n x^n} \right| = \rho.$$

This expression can be rewritten as

$$|x| \lim_{n \to \infty} \left| \frac{a_{n+1}}{a_n} \right| = \rho.$$

Recall that the ratio test states that a series is absolutely convergent for $\rho < 1$. So a power series is absolutely convergent when

$$|x| \lim_{n \to \infty} \left| \frac{a_{n+1}}{a_n} \right| < 1.$$

Rewriting this expression, we find that the interval of convergence for the power series is given by

$$|x| < \lim_{n \to \infty} \left| \frac{a_n}{a_{n+1}} \right|.$$

The limit term on the right-hand side of this expression is traditionally set equal to r:

$$\lim_{n \to \infty} \left| \frac{a_{n+1}}{a_n} \right| = r,$$

which is referred to as the **radius of convergence**. We have now determined that the given expression for a power series is absolutely convergent in the interval $-r < x < r$. Note that for the power series representation

$$f(x) = \sum_{n=0}^{\infty} a_n x^n,$$

the interval of convergence $-r < x < r$ is centered around the origin ($x = 0$).

Recall from Section 2.3.5 that the ratio test is indeterminate for $\rho = 1$, and this condition will occur at each endpoint of an interval. This means that the ratio test cannot determine whether the series is divergent or convergent at these points. Convergence at the endpoints $x = r$ and $x = -r$ will always have to be tested separately and by some method other than the ratio test. To test a series at the endpoints of an interval, substitute the values of the endpoint ($\pm r$) into the series expansion and test the resulting series (again, by some method other than the ratio test).

If we apply the ratio test to the power series expression

$$f(x) = \sum_{n=0}^{\infty} a_n (x - c)^n,$$

we will find that the series is absolutely convergent in the interval $-r < (x - c) < r$. That is, the interval of convergence is now centered around the point $x = c$ rather than around the origin. Once again, the convergence of the series at the endpoints $(c - r)$ and $(c + r)$ must be checked separately and by some method other than the ratio test. The point of this discussion has been to draw attention to the fact that the expression

$$f(x) = a_0 + a_1 x + a_2 x^2 + \cdots = \sum_{n=0}^{\infty} a_n x^n$$

for a power series is an expansion of a function around the origin, while the power series expression

$$f(x) = a_0 + a_1(x-c) + a_2(x-c)^2 + \cdots = \sum_{n=0}^{\infty} a_n(x-c)^n$$

is a series expansion of a function around the point $x = c$.

As an example, consider finding the interval of convergence for the power series

$$\sum_{n=0}^{\infty} \frac{x^n}{n}.$$

Applying the ratio test, we find that

$$\lim_{n \to \infty} \left| \frac{nx^{n+1}}{(n+1)x^n} \right| = \rho$$

or

$$|x| \lim_{n \to \infty} \left| \frac{a_{n+1}}{a_n} \right| = \rho.$$

A series is absolutely convergent when $\rho < 1$, so the given power series is convergent when

$$|x| \lim_{n \to \infty} \left| \frac{n}{(n+1)} \right| < 1.$$

As $n \to \infty$, this limit term approaches 1:

$$\lim_{n \to \infty} \left| \frac{n}{(n+1)} \right| \to 1,$$

so the power series converges for

$$|x| < 1.$$

We now know that the power series converges for values of x *strictly between* 1 and -1, but we do not know how the series behaves at the endpoints 1 and -1. To test the series at the endpoints, we substitute the value of each endpoint into the series and retest by some other method. At the endpoint $x = 1$, the power series takes the form

$$\sum_{n=0}^{\infty} \frac{x^n}{n} = \sum_{n=0}^{\infty} \frac{1}{n}.$$

Figure 3.1 The interval of convergence.

We recognize this series as a harmonic series, which is divergent (by the integral test). So the given power series diverges at $x = 1$.

At the endpoint $x = -1$, the power series takes the form

$$\sum_{n=0}^{\infty} \frac{x^n}{n} = \sum_{n=0}^{\infty} \frac{1}{n} = -1 + \frac{1}{2} - \frac{1}{3} + \frac{1}{4} + \cdots.$$

This is an alternating harmonic series, which is convergent (by the alternating series test). Pulling all of this together, we conclude that the power series

$$\sum_{n=0}^{\infty} \frac{x^n}{n}$$

is convergent for $-1 \le x < 1$.

Intervals of convergence can also be expressed using **interval notation**. In interval notation, brackets and parentheses are used to indicate whether an endpoint is **included** or **excluded** from the interval, i.e., whether it is **closed** or **open**. Using interval notation, the interval of convergence $-1 \le x < 1$ is written as $[-1, 1)$. The bracket indicates that -1 is included in the interval and therefore that the interval is closed at $x = -1$, while the parenthesis indicates that 1 is excluded from the interval and therefore that the interval is open at $x = 1$. The interval of convergence $[-1, 1)$ can be visually represented as shown in Figure 3.1.

Power series of the form

$$\sum_{n=0}^{\infty} a_n x^n$$

have very simple convergence properties, as only one of the following three statements is true of such a series:

i *The series is absolutely convergent for all x,*
ii *The series is convergent only for $x = 0$, or*
iii *The series is absolutely convergent for $|x| < r$ and divergent for $|x| > r$, where r is a positive number.*

Similarly, power series of the form

$$\sum_{n=0}^{\infty} a_n(x - c)^n$$

can also have only one of the following three properties:

i *The series is absolutely convergent for all x,*
ii *The series is convergent only when (x − c) = 0, i.e., when x = c, or*
iii *The series is absolutely convergent when |x − c| < r and is divergent when | x − c| > r, where r is a positive number.*

Many mathematical functions have a **power series representation** or **expansion**. If a function has a power series expansion (not all do), then it is unique, and the power series will converge to that function (within the interval of convergence). Functions defined in this manner have properties similar to those of polynomials and can be treated as such. That is, they can be added, subtracted, multiplied, differentiated, and so forth. The usefulness of power series representations lies in the fact that the series representation gives us a simple technique for evaluating functions, their integrals, and their derivatives. For example, if the function $f(x)$ has the power series expansion

$$f(x) = a_0 + a_1 x + a_2 x^2 + a_3 x^3 + \cdots,$$

then the integral of $f(x)$ can be found by integrating the power series term by term:

$$\int f(x) = a_0 x + \frac{a_1 x^2}{2} + \frac{a_2 x^3}{3} + \cdots$$

The power series that results from the term-by-term integration will converge to the integral of the original function (within the interval of convergence of the original series), although this is not necessarily true at the endpoints of the interval. This can be a powerful method for handling difficult integrals. Even if the integration of the function $f(x)$ is difficult, the integration of the individual terms of its powers series expansion will always be simple, as this only involves integrating integer powers of x. If we need to find the derivative of $f(x)$, we can simply differentiate the power series of $f(x)$ term by term to produce

$$f'(x) = a_1 + 2a_2 x + 3a_2 x^2 + \cdots$$

The resulting series will converge to the derivative of the original function within the original interval of convergence. Once again, this is not necessarily true at the endpoints of the interval. And if we want to evaluate the function $f(x)$

at a point $x = c$ (within the interval of convergence), we can sum the series expression

$$f(c) = a_0 + a_1 c + a_2 c^2 + a_3 c^3 + \cdots = \sum_{n=0}^{\infty} a_n c^n.$$

The general properties of power series are listed below.

Theorem 3.1 *If a function has a power series expansion or representation, it is unique.*

Theorem 3.2 *Convergent series can be added, subtracted, or multiplied, and the resulting series is convergent at least within the interval of convergence common to all the series.*

Theorem 3.3 *Convergent power series can be divided as long as there is no division by zero, unless a zero occurring in the denominator series is canceled by a zero occurring in the numerator series. The interval of convergence for the resulting series will need to be determined by the ratio test.*

Theorem 3.4 *A power series can be substituted into another power series provided that the interval of convergence of the substitute series lies within the interval of convergence of the original series.*

Theorem 3.5 *A power series can be differentiated or integrated term by term, and the resulting series will converge to the derivative or integral of the function represented by the original series within the interval of convergence of the original series.*

3.3 Power Series Expansions of Functions

In Section 3.2, we discussed the properties of power series and how those properties make it easy to manipulate and combine power series. However, we have not yet indicated how to actually find the power series expansion of a function (if it exists). One direct technique for doing this (which is not very difficult) is to perform a **Taylor series expansion** of the function around the point of interest.

The first step in this technique is to *assume* that there is a power series representation of the function around the point of interest at $x = c$, so that $f(x)$ can be written as

$$f(x) = a_0 + a_1(x - c) + a_2(x - c)^2 + a_3(x - c)^3 + \cdots$$

We now need to determine the value of the coefficients a_n, although this may at first sight seem daunting for this expression. However, the task is not difficult if we recognize that in expanding a function, we are trying to precisely or fully describe the function at the point of interest. In order to fully describe a function at a single point, we need to state the value of the function at that point as well as its slope and the rate of change of the slope, the rate of change of the rate of change of the slope, ... That is, in order to fully describe a function at a single point, we need to specify the value of the function as well as the value of all its derivatives: $f(c), f'(c), f''(c), f'''(c)$, and so forth.

In expanding the function $f(x)$ around the point $x = c$, we need to evaluate both the function and the series expression at that point. Setting $x = c$ in both the function and the series expansion results in the expression

$$f(c) = a_0 + a_1 \times 0 + a_2 \times 0^2 + a_3 \times 0^3 + \cdots$$

Note that all of the terms in the series except for the first term are exactly zero. In fact, we have now determined that the value of the coefficient a_0 is simply the value of the function $f(x)$ at $x = c$, i.e.,

$$f(c) = a_0.$$

Plugging the value of a_0 into the series expansion, we get the expression

$$f(x) = f(c) + a_1(x - c) + a_2(x - c)^2 + a_3(x - c)^3 + \cdots$$

Having found the value of the function at the point of interest, we now need to determine the value for the slope of the function (the first derivative) at the point of interest. Taking the first derivative of both the function and the series expansion gives us

$$f'(x) = a_1 + 2a_2(x - c) + 3(x - c)^2 + \cdots + na_n(x - c)^{n-1} + \cdots$$

Evaluating the expression for the first derivative at the point $x = c$, we find that

$$f'(c) = a_1 + 2a_2 \times 0 + 3a_3 \times 0^3 + \cdots$$

Once again, all the terms in the expression evaluate to zero, except for the first term, which is the coefficient a_1:

$$f'(c) = a_1.$$

Substituting the value for a_1 into our series expansion, we find that

$$f(x) = f(c) + f'(c)(x - c) + a_2(x - c)^2 + a_3(x - c)^3 + \cdots$$

Having determined the value of the function and its slope at the point $x = c$, we now need to determine the rate of change of the slope, i.e. the second derivative, at the point of interest. Taking the second derivative of both the function and its series representation we get

$$f''(x) = 2a_2 + 6a_3(x - c) + \cdots + n(n - 1)a_n(x - c)^{n-2} + \cdots$$

When evaluating the second derivative at $x = c$, we find that

$$f''(x) = 2a_2 + 6a_3 \times 0 + \cdots + n(n - 1)a_n \times 0^{n-2} + \cdots$$

so that

$$f''(c) = 2a_2.$$

Continuing in this fashion, we will find that the nth derivative of the function is related to the nth coefficient of the Taylor series expansion by the formula

$$f^{(n)}(c) = n!a_n.$$

Rearranging this formula, we obtain the following expression for the nth coefficient a_n of the series expansion:

$$a_n = \frac{f^{(n)}(c)}{n!}.$$

So if a function $f(x)$ has a power series representation of the form

$$f(x) = \sum_{n=0}^{\infty} a_n(x - c)^n, \tag{3.5}$$

the coefficients a_n of the series are given by

$$a_n = \frac{f^{(n)}(c)}{n!}. \tag{3.6}$$

Alternatively, we could write the series expansion as

$$f(x) = f(c) + f'(c)(x - c) + \frac{f''(c)}{2!}(x - c)^2 + \frac{f'''(c)}{3!}(x - c)^3 + \cdots$$

$$+ \frac{f^{(n)}(c)}{n!}(x - c)^{n+3} + \cdots,$$

which is the expression for a **Taylor series expansion of $f(x)$ around the point** $x = c$.

As an example, we will find the power series expansion for $\sin x$ around the point $x = \pi$. The first step is to assume that there is a power series expansion of the form

$$\sin x = a_0 + a_1(x - \pi) + a_2(x - \pi)^2 + a_3(x - \pi)^3 + \cdots$$

To fully describe $\sin x$ around the point $x = \pi$, we need to specify the value of the function and all of its derivatives at that point.

To determine the value of a_0, set $x = \pi$ in both the function and the series. The resulting expression looks like

$$\sin x = a_0 + a_1(0) + a_2(0)^2 + a_3(0)^3 + \cdots$$

Since $\sin \pi = 0$, it follows that $a_0 = 0$, so the series expansion has the form

$$\sin x = a_1(x - \pi) + a_2(x - \pi)^2 + a_3(x - \pi)^3 + \cdots$$

We now need to determine the value of the first derivative at $x = \pi$. Taking the derivative of both the function and the series, we find that

$$\cos x = a_1 + 2a_2(x - \pi) + 3a_3(x - \pi)^2 + \cdots + na_n(x - \pi)^{n-1} + \cdots$$

Evaluating this expression at $x = \pi$, we get

$$\cos \pi = a_1 + 2a_2 \times 0 + 3a_3 \times 0^2 + \cdots + na_n \times 0^{n-1} + \cdots$$

The value of $\cos \pi$ is -1, so

$$\cos \pi = -1 = a_1$$

Substituting the value of a_1 into the series expansion for $\sin(x)$, we find that

$$\sin x = -(x - \pi) + a_2(x - \pi)^2 + a_3(x - \pi)^3 + \cdots + a_n(x - \pi)^n + \cdots$$

To specify the value of the second derivative, we take the second derivative of both the function and the series and set $x = \pi$ to find that

$$-\sin \pi = 0 = a_2.$$

If we continue taking successive derivatives of both the function and the series expansion, evaluating at $x = \pi$ each time, eventually we will arrive at the series expansion for $\sin x$ around the point $x = \pi$:

$$\sin x = -(x - \pi) + \frac{(x - \pi)^3}{3!} - \frac{(x - \pi)^5}{5!} + \cdots$$

While a Taylor series expansion is a straightforward, direct method for determining the coefficients of a power series, it can be labor intensive.

Furthermore, the Taylor series expansion can be impractical in situations where the successive or higher-order derivatives of the function are not easy to compute. For example, consider expanding the function $f(x) = \arctan x$ in terms of a Taylor series. We would quickly discover that successive differentiation of the function becomes onerous:

$$f(x) = \arctan x,$$
$$f'(x) = \frac{1}{1-x^2},$$
$$f''(x) = \frac{2x}{(1-x^2)^2},$$
$$f'''(x) = \frac{2}{(1-x^2)^2} - \frac{8x^2}{(1-x^2)^3}.$$

Note that Taylor series expansions require successive differentiation of the function. However, differentiation is not defined at discontinuities. So if a function has discontinuities, it is not possible to find the derivative of the function at the discontinuities. This is why functions with discontinuities, or "sharp corners," cannot be expanded in terms of a power series at the discontinuity. We will have to wait for the introduction of additional methods (e.g., Fourier series expansions) to be able to expand functions with discontinuities.

One final point is that a Taylor series expansion performed around the origin ($c = 0$) is referred to as a **Maclaurin series expansion**. The Maclaurin series expansion for a function $f(x)$ is given by

$$f(x) = f(0) + f'(0) + f''(0)\frac{x^2}{2!} + f'''(0)\frac{x^3}{3!} + \cdots + f^{(n)}(0)\frac{x^n}{n!} + \cdots \quad (3.7)$$

This expression is powerful in that it explicitly demonstrates that a power series representation of a function is simply a full description of the function's behavior at the point of interest (in this case, the origin). Reading the terms of the expression from left to right, the behavior of the function $f(x)$ at the origin is described by its value $f(0)$ at the origin, its slope $f'(0)$ at the origin, the rate of change $f''(0)$ of its slope at the origin, and so on. The Maclaurin series expansion for a function is useful, and in many cases it is possible to obtain the Taylor series expansion from the Maclaurin series with a simple substitution, thereby saving ourselves from unnecessary toil.

3.4 Other Methods for Constructing Power Series Expansions

Using a Taylor series expansion to find the power series representation of an unfamiliar function can be labor intensive and might not always be the best choice. Frequently, there are clever, simpler alternatives for finding the power series expansion of a function. We know from Section 3.2 that the power series representation of a function is unique, i.e., there is one and only one such series expansion. This implies that whatever clever or simpler alternative methods we might use to find a power series, it will be identical to the series would have been found using the labor-intensive Taylor series expansion.

In many cases, it is possible to manipulate a known series expansion using substitution, addition, subtraction, multiplication, division, differentiation, or integration to construct the power series representation for a function. The key to this method is to memorize the power series for a few fundamental functions. The easiest power series to memorize is the geometric series for $1/(1-x)$ and the series expansions for the functions $\ln(x+1)$, $(1+x)^p$, e^x, $\sin x$, and $\cos x$:

$$\frac{1}{1-x} = 1 + x + x^2 + x^3 + \cdots = \sum_{n=0}^{\infty} x^n, \qquad \text{for} |x| < 1; \qquad (3.7)$$

$$\ln(1+x) = x - \frac{x^2}{2} + \frac{x^3}{3} - \frac{x^4}{4} + \cdots = \sum_{n=0}^{\infty} \frac{(-1)^{n+1} x^n}{n!}, \qquad \text{for} -1 < x \le 1;$$

$$(3.8)$$

$$(1+x)^p = 1 + px + \frac{p(p-1)}{2!} x^2 + \frac{p(p-1)(p-2)}{3!} x^3 + \cdots$$

$$= \sum_{n=0}^{\infty} \binom{p}{n} x^n, \text{for} |x| < 1; \qquad (3.9)$$

$$e^x = 1 + x + \frac{x^2}{2!} + \frac{x^3}{3!} + \cdots = \sum_{n=0}^{\infty} \frac{x^n}{n!}, \qquad \text{for all } x; \qquad (3.10)$$

$$\sin x = x - \frac{x^3}{3!} + \frac{x^5}{5!} - \frac{x^7}{7!} + \cdots = \sum_{n=0}^{\infty} \frac{(-1)^n x^{2n+1}}{(2n+1)!}, \qquad \text{for all } x; \qquad (3.11)$$

$$\cos x = 1 - \frac{x^2}{2!} + \frac{x^4}{4!} - \frac{x^6}{6!} + \cdots = \sum_{n=0}^{\infty} \frac{(-1)^n x^{2n}}{(2n)!}, \quad \text{for all } x. \quad (3.12)$$

We will make use of these power series to demonstrate how a known series can be manipulated to find the power series representation for an unfamiliar function.

3.4.1 Substitution

Replacing or substituting the variable in a known series with a different variable is one of the most straightforward ways to create another power series. For example, we can replace the variable x with a function of x, such as $-x$, $1/x$, or x^2, to construct another series. Consider finding the power series representation for the function

$$\frac{1}{1+x} = \sum_{n=0}^{\infty} a_n x^n.$$

Rather than resorting to a Taylor series expansion, it pays to note that this expression is similar to the expression for the sum of the geometric series:

$$\frac{1}{1-x} = 1 + x + x^2 + x^3 + \cdots = \sum_{n=0}^{\infty} x^n.$$

Substituting $-x$ for x in this series expansion results in the expression $1/(1+x)$, which is now recognizable as the sum of a geometric series in powers of $-x$:

$$\frac{1}{1+x} = \frac{1}{1-(-x)} 1 - x + x^2 - x^3 + \cdots = \sum_{n=0}^{\infty} a_n(-x)^n.$$

In many cases, it is possible to find the Taylor series expression for a function by simply making a substitution in the known Maclaurin series expression. To see how this is done, suppose we need the Taylor series expression for sin x around the point $x = \pi$. The Maclaurin series expansion for sin x is

$$\sin x = x - \frac{x^3}{3!} + \frac{x^5}{5!} - \frac{x^7}{7!} + \cdots = \sum_{n=0}^{\infty} \frac{(-1)^n x^{2n+1}}{(2n+1)!}.$$

This is the expansion of sin x around the origin, or in other words, an expansion in powers of x, or equivalently in powers of $x - 0$, i.e.,

$$sin(x-0) = (x-0) - \frac{(x-0)^3}{3!} + \frac{(x-0)^5}{5!} - \frac{(x-0)^7}{7!} + \cdots$$

To find the expansion for $\sin x$ around the point $x = \pi$, or in of powers of $x - \pi$, we simply substitute $x - \pi$ for $x - 0$ in the preceding expression to get

$$-\sin(x-\pi) = -\sin x = (x-\pi) - \frac{(x-\pi)^3}{3!} + \frac{(x-\pi)^5}{5!} - \cdots$$

$$= \sum_{n=0}^{\infty} \frac{(-1)^n (x-\pi)^{2n+1}}{(2n+1)!}.$$

Using the trigonometric identity for $\sin(x-\pi)$, we get

$$\sin x = -(x-\pi) + \frac{(x-\pi)^3}{3!} - \frac{(x-\pi)^5}{5!} - \cdots = \sum_{n=0}^{\infty} \frac{(-1)^{n+1}(x-\pi)^{2n+1}}{(2n+1)!}.$$

Note that this is the same expression we found in Section 3.3, where we used the Taylor series expansion to find the power series representation for $\sin x$ around the point $x = \pi$.

It is even possible to replace the variable of a series with another series. Such a substitution is legitimate provided that the values of the substituted series are within the interval of convergence of the original series. For example, to find the series expansion for $e^{\sin x}$, we simply replace x in the series expansion for e^x with $\sin x$:

$$e^{\sin x} = 1 + \sin x + \frac{(\sin x)^2}{2!} - \frac{(\sin x)^3}{3!} + \cdots = \sum_{n=0}^{\infty} \frac{(\sin x)^n}{n!}.$$

This is a legitimate substitution as the values of $\sin x$ (from 1 to -1) are within the interval of convergence (all values of x) for the power series for e^x. We can even go a step further, and replace $\sin x$ with its series representation to obtain

$$e^{\sin x} = 1 + \left(x - \frac{x^3}{3!} + \frac{x^5}{5!} - \cdots\right) + \frac{(x - \frac{x^3}{3!} + \frac{x^5}{5!} - \cdots)^2}{2!} + \frac{(x - \frac{x^3}{3!} + \frac{x^5}{5!} - \cdots)^3}{3!} + \cdots$$

Collecting like powers leads to the result

$$e^{\sin x} = 1 + x + \frac{x^2}{2!} + \frac{x^4}{4!} + 2\frac{x^5}{5!} + \cdots, \text{ for all } x.$$

3.4.2 Multiplication

It is possible to create a series by multiplying a series by a polynomial or another series. For example, to find the series representation for the function $(1 + x^2)$ $(\sin x)$, we can simply multiply the series expression for $\sin x$ by $(1 + x^2)$ to get

$$(1 + x^2)\sin x = (1 + x^2)\left(x - \frac{x^3}{3!} + \frac{x^5}{5!} - \frac{x^7}{7!} + \cdots\right).$$

Completing the multiplication and collecting like terms, we find that

$$(1 + x^2)\sin x = x + x^3\left(\frac{1}{1!} - \frac{1}{3!}\right) - x^5\left(\frac{1}{3!} - \frac{1}{5!}\right) + x^7\left(\frac{1}{5!} - \frac{1}{7!}\right) + \cdots$$

This process for finding the power series expansion for an unfamiliar function was far easier than resorting to the Taylor series method, which would have required taking successive derivatives of $(1 + x^2)$ $(\sin x)$.

While the previous example involved multiplication of a series by a polynomial, it is also possible to find a series expansion by multiplying one series by another. For example, consider finding the power series representation for $2 / (x^2 - 1)$. Notice that the function can be written as a product of three terms:

$$\frac{2}{x^2 - 1} = 2 \times \frac{1}{x - 1} \times \frac{1}{1 + x}.$$

The second term in the product, $1 / (x - 1)$, is recognizable as the sum of a geometric series, i.e.,

$$\frac{1}{x - 1} = \frac{-1}{1 - x} = -1 - x - x^2 - x^3 - \cdots$$

while the third term, $1 / (1 + x)$, is also the sum of a geometric series. This can be seen through a substitution like the one we did in Section 3.4.1. Substituting $-x$ for x in the expression for the geometric sum $1 / (1 - x)$, we find that

$$\frac{1}{1 + x} = \frac{1}{1 - (-x)} = 1 - x + x^2 - x^3 + \cdots$$

The function $2 / (x^2 - 1)$ can be seen to be the product of a constant and two geometric series:

$$\frac{2}{x^2 - 1} 2 \times \frac{-1}{1 - x} \times \frac{1}{1 + x} = 2(-1 - x - x^2 - \cdots)(1 - x + x^2 - x^3 + \cdots).$$

Completing the multiplication and combining like powers, we find that

$$\frac{2}{x^2 - 1} = -2(1 + x^2 + x^4 + x^6 + \cdots).$$

If the power series are expressed using sigma notation, it is possible to complete the multiplication as follows. If

$$f(x) = \sum_{n=0}^{\infty} a_n x^n$$

and

$$g(x) = \sum_{n=0}^{\infty} b_n x^n,$$

then the product of the two series is

$$f(x)g(x) = \sum_{n=0}^{\infty} a_n x^n \sum_{n=0}^{\infty} b_n x^n = \sum_{n=0}^{\infty} c_n x^n,$$

where

$$c_n = a_0 b_1 + a_1 b_{n-1} + \cdots + a_n b_0 = \sum_{k=0}^{n} a_k b_{k-1} = \sum_{k=0}^{n} a_{k-1} b_k.$$

3.4.3 Division

Series expansions can also be obtained by dividing a series by a polynomial or another series. To obtain the series for $(1 / x) \sin x$, we simply divide the power series expression for $\sin x$ by x:

$$\frac{1}{x} \sin x = \frac{1}{x}(x - \frac{x^3}{3!} + \frac{x^5}{5!} - \frac{x^7}{7!} + \cdots) = 1 - \frac{x^2}{3!} + \frac{x^4}{5!} - \frac{x^6}{7!} + \cdots$$

This process is clearly much simpler than finding the Taylor series expansion, which would have required taking successive derivatives of $(1 / x) \sin x$.

It is also possible to divide a series by another series. For example, the series expansion for $\tan x$ can be found by dividing the series for $\sin x$ by the series for $\cos x$:

$$\tan x = \frac{\sin x}{\cos x} = \frac{(x - \frac{x^3}{3!} + \frac{x^5}{5!} - \frac{x^7}{7!} + \cdots)}{(1 - \frac{x^2}{2!} + \frac{x^4}{4!} - \frac{x^6}{6!} + \cdots)}.$$

The division can be accomplished using long division:

$$1 - \frac{x^2}{2!} + \frac{x^4}{4!} - \frac{x^6}{6!} + \cdots \overline{\Big)\ x - \frac{x^3}{3!} + \frac{x^5}{5!} - \frac{x^7}{7!} + \cdots}$$

$$x + \frac{x}{3!} + \frac{2}{15}x^5 + \cdots$$

$$\underline{x - \frac{x^3}{2!} + \frac{x^5}{4!} - \cdots}$$

$$\frac{x^3}{3} - \frac{x^5}{30} + \cdots$$

$$\underline{\frac{x^3}{3} - \frac{x^5}{6} + \cdots}$$

$$\frac{2x^5}{15} + \cdots$$

$$\vdots$$

3.4.4 Differentiation

It is often possible to integrate or differentiate a known series to obtain the series expansion for another function. For example, if we know the power series representation for the function $f(x)$ and need the power series expansion for the function $df(x)\,/\,dx$, we can simply differentiate the series representation of $f(x)$. The term-by-term differentiation of the power series expansion of a function will result in a series expansion that converges to the derivative of the original function within the interval of convergence of the original function, although not necessarily at the endpoints of the interval. So we would need to retest for convergence at the endpoints.

As an example, consider the power series expansion for the function $1\,/\,(1-x)^2$:

$$\frac{1}{(1-x)^2} = \sum_{n=0}^{\infty} a_n x^n.$$

Rather than immediately attempting a Taylor series expansion of the function, it is worth considering the similarity between the expression $1\,/\,(1-x)^2$ and the expression $1\,/\,(1-x)$, which is the sum of a geometric series. It might occur to you that $1\,/\,(1-x)^2$ is simply the product of $1\,/\,(1-x)$ with itself:

$$\frac{1}{(1-x)^2} = \frac{1}{(1-x)} \times \frac{1}{(1-x)}.$$

Each factor $1 / (1 - x)$ can be written as the geometrics series $1 + x + x^2 + x^3 + \cdots$, so that the power series expansion for $1 / (1 - x)^2$ is the product of two geometric series:

$$\frac{1}{(1-x)^2} = (1 + x + x^2 + x^3 + \cdots) \times (1 + x + x^2 + x^3 + \cdots).$$

You could work through the multiplications and collect like terms to find that

$$\frac{1}{(1-x)^2} = (1 + 2x + 3x^2 + 4x^3 + \cdots).$$

If you find grinding out the multiplication of two infinite polynomials and collecting like terms distasteful, take heart. In this example, there is a quicker way to find the power series expansion using differentiation. Notice that $1 / (1 - x)^2$ is the derivative of $1 / (1 - x)$:

$$\frac{d}{dx}\left(\frac{1}{1-x}\right) = \frac{1}{(1-x)^2}.$$

Therefore, we could have found the power series of interest by simply taking the derivative of the geometric series expansion for $1 / (1 - x)$:

$$\frac{1}{(1-x)^2} = \frac{d}{dx}\left(\frac{1}{1-x}\right) = \frac{d}{dx}(1 + x + x^2 + x^3 + \cdots)$$

$$= (1 + 2x + 3x^2 + 4x^3 + \cdots).$$

The function $1 / (1 - x)$ converges for x in the interval $(-1, 1)$, the interval of convergence for the geometrics series, and, therefore, the function $1 / (1 - x)^2$ also converges within the same interval. However, the convergence behavior at the endpoints needs to be checked separately. Substituting the values 1 or -1 for the values of x at the endpoints, we find that the series diverges at both endpoints, and therefore the interval of convergence is open, i.e., it is $(-1, 1)$.

3.4.5 Integration

It is possible to find the series representation of a function by integrating a known power series. For example, if we need the series expansion of a function $g(x)$ and can express $g(x)$ as the integral of a function $f(x)$ whose series expansion is known, we can simply integrate the series expression for $f(x)$ to find the series for $g(x)$. The term-by-term integration of a series will result in a series that converges to the integral of the original series, at least within the interval of convergence of the original series. This will not necessarily be true at the endpoints of the interval, so once again convergence will need to be tested separately at these points

For example, if we need the power series expansion for the function $\ln(1-x)$, the function $\ln(x)$ can be expressed as the elementary integral

$$\ln x = \int \frac{1}{x} dx.$$

Substituting $1-x$ for x in this equation results in the expression

$$\ln (x-1) = -\int \frac{dx}{1-x},$$

where the integrand $1/(1-x)$ has a known power series expansion (a geometric series). By making a substitution in an elementary integral, we have managed to express the function $\ln(1-x)$ as the integral of a known series expansion. So

$$\ln(x-1) = -\int \frac{dx}{1-x} = -\int (1+x+x^2+x^3+\cdots)dx.$$

The integration is straightforward, as it only involves integer powers of x, and results in the expression

$$\ln (1-x) = -\left(x+\frac{x^2}{2}+\frac{x^3}{3}+\frac{x^4}{4}+\cdots\right).$$

This expression converges at least within the interval of convergence of the original geometric series, $(-1, 1)$. However, the condition for convergence needs to be checked separately at the endpoints. Evaluating the series at the endpoint $x = 1$ results in a negative harmonic series, and therefore it is divergent. Evaluating the series at the endpoint $x = -1$ results in an alternating harmonic series, which is convergent. Therefore, in interval notation, the interval of convergence for the power series expansion of $\ln(1-x)$ is $[-1, 1)$.

3.4.6 Addition, Subtraction, Partial Fractions, and Combined Methods

We might be able to obtain the series expression for an unfamiliar function simply by splitting the function up until it is recognizable as the sum of two (or more) familiar expressions. The familiar expressions can then be written out in their series expansions, and the series added or subtracted (term by term) to obtain the series representation for the unfamiliar function.

In Section 3.4.2, we found the power series representation for the function $2 / (x^2 - 1)$ by recognizing it as the product of the geometric series $2 / (x - 1)$ and $1 / (1 + x)$. It is also possible to obtain the series expansion for $2 / (x^2 - 1)$ by making use of **partial fraction decomposition** to split the function up so that it is recognizable as the sum of more familiar expressions [9]. The method of partial fraction decomposition is a useful tool as it is applicable to rational expressions. A **rational function**, or a **rational expression**, is the quotient of two polynomials:

$$f(x) = \frac{g(x)}{h(x)}.$$

If the degree of the polynomial $g(x)$ is less than the degree of the polynomial $h(x)$, then in theory it is always possible to write a rational expression for $f(x)$ as the sum of rational functions whose denominators are powers of polynomials with a degree not greater than two [2, 10]. That is, it is theoretically possible to express $f(x)$ as

$$f(x) = \frac{g(x)}{h(x)} = \frac{A_1}{(x - q_1)^{n_1}} + \frac{A_2}{(x - q_2)^{n_2}} + \frac{A_3}{(x - q_3)^{n_3}} + \cdots \qquad (3.13)$$

The individual terms $A_n / (x - q_n)^n$ on the right-hand side of this equation can then be related to an expression for the sum of a geometric series or its derivative.

Consider the previous example function, $2 / (x^2 - 1)$. It is easy to verify that its partial fraction decomposition is given by

$$\frac{2}{(x^2 - 1)} = \frac{1}{(x - 1)} + \frac{-1}{(x + 1)}.$$

The two terms on the right-hand side of this expression can be rewritten so that it is obvious that each is the sum of a geometric series:

$$\frac{2}{(x^2 - 1)} = \frac{-1}{(1 - x)} + \frac{-1}{(1 - (-x))}.$$

The terms on the right-hand side can now be written out as their geometric series expansions:

$$\frac{2}{(x^2 - 1)} = -(1 + x + x^2 + x^3 + \cdots) + (-1 + x - x^2 + x^3 - x^4 + \cdots).$$

Adding the series together term by term results in the expression

$$\frac{2}{(x^2 - 1)} = (-2 - 2x^2 - 2x^4 - \cdots) = -2(1 + x + x^2 + x^3 + \cdots).$$

This is identical to the power series found in Section 3.4.2, which was obtained by multiplying two series together.

3.5 Accuracy of Series Approximations

The previous subsections detailed various methods by which the infinite power series representation of a function can be obtained. In each case, the series expansion was an exact representation of the function (within the interval of convergence of the series). Such representations are exact but only if one fully sums the infinite number of terms in the series. In many applied problems, it is not necessary to sum all the terms of a series. Typically, it is only necessary to sum enough terms to get the accuracy required for the application or the nature of the problem. In such cases, it is possible to truncate the series and use a finite number of terms to approximate the value of the function of interest. We briefly touched upon this subject at the end of Section 2.2, where we introduced the concept of a remainder. In Chapter 2, the remainder R_n, or the error, was defined as the sum of all the neglected terms of an infinite series. That is, the remainder is a measure of how accurately the sum of a finite number of terms approximates the sum of an infinite number of terms.

For a Taylor series expansion, the **remainder** $R_n(x)$ is defined as the difference between the true value of the function $f(x)$ and the sum of the first $n + 1$ terms of the function's series expansion:

$$R_n(x) = f(x) - \left[f(c) + (x - c)f'(c) + (x - c)^2 \frac{f''(x)}{2!} + \cdots + (x - c)^n \frac{f^{(n)}(x)}{n!} \right].$$

$$(3.14)$$

The first $n + 1$ terms of a Taylor series form a polynomial referred to as a **Taylor polynomial**. In the preceding expression, the complete Taylor series expansion was replaced with the **nth Taylor polynomial**. While it will not be proven, the following **generalized mean value theorem** can be used to derive a compact expression for the remainder $R_n(x)$:

$$R_n(x) = (x - c)^{n+1} \frac{f^{(n+1)}(\varepsilon)}{(n + 1)!}.$$

The variable ε is some yet-to-be determined point between x and c. The location of ε is found by setting the derivative of $R_n(x)$ equal to zero and solving for ε, i.e., we find the local extrema of the remainder. The real usefulness of this expression for the reminder $R_n(x)$ becomes clear when we realize that the given expression is simply the first neglected term of the Taylor series. *That is, if we approximate a function by only considering the first n terms of its series expansion, then the remainder or estimated error associated with the approximation scales as the first neglected term of the series.*

The estimated error associated with truncating an alternating series is easy to memorize: *If a convergent alternating series is truncated, then the absolute value of the error associated with the truncation is less than the absolute value of the first neglected term.* Put another way, if

$$\sum_{n=0}^{\infty} a_n = Sum$$

is a convergent alternating series such that

$$|a_{n+1}| < |a_n|$$

and

$$\lim_{n \to \infty} a_n = 0,$$

then

$$|Sum - (a_1 + a_2 + a_3 + \cdots + a_n)| \leq |a_{n+1}|.$$

For example, consider the convergent alternating geometric series

$$1 - \frac{1}{3} + \frac{1}{9} - \frac{1}{27} + \frac{1}{81} - \cdots = \sum_{n=0}^{\infty} 1 \left(\frac{-1}{3} \right)^n = \frac{1}{1 + \frac{1}{3}} = \frac{3}{4} = 0.75.$$

The sum of the first four terms of the series,

$$1 - \frac{1}{3} + \frac{1}{9} - \frac{1}{27},$$

differs from the sum of the series by about -0.009. The value of the first neglected term of the series, $1/81$, is about 0.012. So the error in approximating the infinite series summation by only summing the first four terms will be less than the absolute value of the first neglected term:

$$\left| Sum - (1 - \frac{1}{3} + \frac{1}{9} + \frac{1}{27}) \right| \cong \left| -0.009 \right| \le \left| \frac{1}{81} \right| \cong 0.012.$$

For convergent series, which do not alternate, the estimated error can be as large as or a few times larger than the first neglected term of the series (and not smaller). For example, consider the non-alternating geometric series

$$1 + \frac{1}{3} + \frac{1}{9} + \frac{1}{27} + \frac{1}{81} + \cdots = \sum_{n=0}^{\infty} 1 \left(\frac{1}{3} \right)^n = \frac{1}{1 - \frac{1}{3}} = \frac{3}{2} = 1.5.$$

In this case, the sum of the first four terms differs (in absolute value) from the sum of the series by about 0.018. This is larger by a factor of two than the error in the previous example. In this case, the error is clearly larger than the first neglected term, 1/81. However, note that the error still has the same order of magnitude as the first neglected term. In other words, *for non-alternating series, the first neglected term of the series has the same order of magnitude as the remainder or estimated error.* In these cases, the estimated error associated with truncating a non-alternating power series can be found in the following way.
    If the power series

$$\sum_{n=0}^{\infty} a_n x^n = Sum$$

converges for $|x| < 1$ and

$$|a_{n+1}| < |a_n| \text{ for } n > N,$$

then the estimated error associated with approximating a non-alternating series by summing the first N terms is given by the following expression: [8]

$$\left| Sum - \left(\sum_{n=0}^{N} a_n x^n \right) \right| \le \left| a_{N+1} x^{N+1} \right| \div (1 - |x|). \tag{3.15}$$

That is, the estimated error will be less than the first neglected term divided by the factor $1 - |x|$. As a concrete example, consider once again the non-alternating geometric series that sums to 1/3 (common ratio $x = 1/3$). The remainder, or the estimated error, in approximating the infinite series by the sum of the first four terms is given by

$$\left| Sum - (1 + \frac{1}{3} + \frac{1}{9} + \frac{1}{27}) \right| \cong \left| -0.018 \right|.$$

This error estimate will be less than the absolute value of the first neglected term, $|1/81|$, divided by the factor $|1 - 1/3|$. Working through the calculation, we find that

$$\left| Sum - (1 + \frac{1}{3} + \frac{1}{9} + \frac{1}{27}) \right| \cong |-0.018| \leq \left| \frac{1}{81} \right| \div (1 - \left| \frac{1}{3} \right|) \cong 0.018.$$

This result supports the statement that the error in estimating a non-alternating series as the sum of the first N terms is given by

$$\left| Sum - \left(\sum_{n=0}^{N} a_n x^n \right) \right| \leq \left| a_{N+1} x^{N+1} \right| \div (1 - |x|). \tag{3.16}$$

3.6 Asymptotic Series Expansions

Up to this point, we have focused on representing functions in terms of convergent series. The series expansion of a function is an expansion whose sequence of partial sums converges to the value of the function as the number of terms n in the partial sum approaches infinity:

$$\lim_{n \to \infty} s_n = \lim_{n \to \infty} (s_1(x) + s_2(x) + s_3(x) + \cdots + s_n(x)) = \sum_{n=1}^{\infty} s_n(x) = f(x).$$

There is another case that we have not considered. This is the case where n, the number of terms in the partial sum, is held finite and the value of x is allowed to approach a limit, perhaps some large value of x. There are situations where a finite partial sum will approach the value of a given function $f(x)$ when

$$\lim_{x \to value} s_n = \lim_{x \to value} (s_1(x) + s_2(x) + s_3(x) + \cdots + s_n(x)) = f(x).$$

The reader should note that the variable in the limit has changed from n to x. Such series occur in the fields of astronomy, astrophysics, and mathematics and are used in the numerical approximation of functions. These types of series are called asymptotic series. Questions concerning the convergence or divergence of an asymptotic series are not central to their application or use. Asymptotic series expansions approximate the value of a function as x approaches a limit, and the number of terms n remains finite. By definition, this is not an infinite series, so that the question of convergence or divergence, the behavior of the nth partial sum as n approaches infinity, does not arise. As we will discover, both convergent and divergent asymptotic series exist.

More formally, an **asymptotic series** is a series expansion in terms of the variable x, which may be convergent or divergent, whose nth partial sum $S_n(x)$ approaches the value of a given function $f(x)$ for a limiting value of x.

A function and its asymptotic expansion are given symbolically by the expression

$$f(x) \underset{x \to \infty}{\sim} \sum_{n=0}^{\infty} u_n(x). \tag{3.17}$$

While the symbol $\sim$ is simple in appearance, its meaning is quite specific (and a mouthful to express). The previous equation is to be read as: *The series $\Sigma u_n(x)$ is the asymptotic expansion of $f(x)$ in the limit as $x \to \infty$ and the number of terms N in the partial sum remains finite.*

Put another way, an infinite series

$$\sum_{n=0}^{\infty} u_n(x)$$

is the asymptotic series expansion for $f(x)$ if, for each fixed N,

$$\lim_{x \to \infty} \left(\left| f(x) - \sum_{n=0}^{N} u_n(x) \right| \div u_N(x) \right) \to 0. \tag{3.18}$$

To better understand this expression, consider the term

$$f(x) - \sum_{n=0}^{N} u_n(x).$$

This is simply the difference between the function $f(x)$ and the Nth partial sum of the series. Recall that we encountered this term at the end of Section 2.2. This term is simply the remainder R_n, or the error, in our series approximation after N terms:

$$f(x) - \sum_{n=0}^{\infty} u_n(x) = R \approx error.$$

With this in mind, the condition

$$\lim_{x \to \infty} \left(\left| f(x) - \sum_{n=0}^{N} u_n(x) \right| \div u_n(x) \right) = \lim_{x \to \infty} \left(\left| R_n(x) \right| \div u_n(x) \right) \to 0 \tag{3.19}$$

simply states (in English) that, *in order for Σu_n(x) to be an asymptotic series expansion of f(x), the ratio between the error, or the remainder of the series, and the last term of the partial sum must tend to zero as x approaches ∞.*

Note that the definition of an asymptotic series that was given earlier makes no reference to the specific form that the series

$$\sum_{n=0}^{\infty} u_n(x)$$

must have in order for it to be considered asymptotic. In reality, an asymptotic series can be any one of the many different types of series expansions. Furthermore, an asymptotic series can be either convergent or divergent. As previously noted, when considering asymptotic series, we are primarily concerned with the behavior of the series for a fixed N as x approaches a limit. This means that the question of convergence (the behavior of the series for fixed x as n approaches infinity) is not central to asymptotic expansions, so that the series may be convergent or divergent.

The next two examples will make use of power series to create asymptotic expansions, one around infinity and the other around the origin. The reader should understand that asymptotic series are not limited to only power series expansions – these examples were chosen for their simplicity and transparency.

The asymptotic expansion of a function $f(x)$ around $x = \infty$ frequently involves series whose terms are inverse powers of x. This can be seen by considering a power series expansion given in terms of powers of $1/x$:

$$u(x) = \sum_{n=0}^{\infty} a_n \left(\frac{1}{x}\right)^n = \sum_{n=0}^{\infty} \frac{a_n}{x^n}.$$

Notice that the consecutive terms $(1/x)^n$ of the series become considerably smaller in the limit as $x \to \infty$. *For some expansions, it may even be possible to neglect higher-order terms altogether and simply truncate the power series at some fixed value of n.* Such a truncation results in an nth partial sum of the form

$$u_n(x) = a_0 + \frac{a_1}{x} + \frac{a_2}{x^2} + \cdots + \frac{a_n}{x^n}.$$

In such cases, $u_n(x)$ will be the asymptotic expansion of $f(x)$ if the condition

$$\lim_{x \to \infty} \left(\left| f(x) - u_n(x) \right| \div \frac{1}{x^n} \right) = \lim_{x \to \infty} \left(\left| R_n(x) - u_n(x) \right| x^n \right) = \lim_{x \to \infty} \left(\left| R_n(x) \right| x^n \right) \to 0$$

is true. In this case, we write

$$f(x) \underset{x \to \infty}{\sim} \sum_{n=0}^{\infty} \frac{a_n}{x^n}.$$

This equation is to be read as: *The given power series is the asymptotic expansion of the function f(x) if, for each fixed N, the ratio of the remainder, or error, to the last term in the summation tends to zero as x approaches infinity.*

The previous example was a hypothetical asymptotic expansion of the function $f(x)$ at infinity. But infinity is not the only point around which one can have an asymptotic expansion. It might be possible to have an asymptotic expansion of a function $f(x)$ around some point other than infinity – for example, the origin. To see this, consider a power series expansion given in terms of powers of x:

$$u(x) = \sum_{n=0}^{\infty} a_n x^n.$$

Notice that in the limit as $x \to 0$, the consecutive terms x^n become progressively smaller in value. *At some point in the approximation, it may become possible to neglect higher-order terms of x* and express the nth partial sum as follows:

$$u_n(x) = a_0 + a_1 x + a_2 x^2 + \cdots + a_n x^n.$$

Once again, we have taken the opportunity to truncate the power series at some fixed value of n while allowing x to approach a limiting value (in this case zero). In cases where the condition

$$\lim_{x \to \infty} \left(\left| f(x) - u_n(x) \right| \div x^n \right) = \lim_{x \to \infty} \left(\left| R_n(x) - u_n(x) \right| \frac{1}{x^n} \right)$$

$$= \lim_{x \to \infty} \left(\left| R_n(x) \right| \frac{1}{x^n} \right) \to 0$$

is fulfilled, we write

$$f(x) \underset{x \to \infty}{\sim} \sum_{n=0}^{\infty} a_n x^n.$$

This should be read as: *The given power series is the asymptotic expansion of the function f(x) if, for each fixed N, the ratio of the remainder, or error, to the last term in the summation tends to zero as x approaches zero.*

It is important to understand that, while asymptotic expansions are useful for making approximations, such approximations are inherently limited in their accuracy. The error associated with an infinite series approximation can be made arbitrarily small simply by including additional terms from the series expansion. In contrast, asymptotic expansions only involve partial sums (finite sums), and when summing over a finite number of terms, the error cannot be made arbitrarily small (except possibly in special cases where $x \to 0$) [9].

In both of the preceding examples, the asymptotic expansion was obtained by neglecting higher-order terms of the infinite series. The justification for truncating infinite series is that *at some point in the approximation, it has become possible to neglect higher-order terms*. In many practical situations, the truncation is justified by the fact that the asymptotic expansion is accurate enough for the intended purpose.

How does one determine the error associated with an asymptotic approximation? The answer to this question is simple and familiar. In general, *the error of an asymptotic approximation is smaller in absolute value and has the same sign as the first neglected term of the partial sum*. In some cases, it may even be possible to reduce the error of the asymptotic expansion by allowing the variable x to approach a limiting value. As concrete examples of asymptotic expansions, we will consider approximations for the value of $\sin x$ and e^x for small values of x, i.e.,

$$\sin x \simeq x$$

and

$$e^x \simeq 1 + x.$$

While these are familiar approximations, you might not think of them as symptotic expansions – but that is exactly what they are. The asymptotic series expansion of $\sin x$ incorporates only the first term of the power series expansion

$$\sin x = x - \frac{x^3}{3!} + \frac{x^5}{5!} - \cdots$$

in the limit as x approaches zero, while the asymptotic series expansion of the exponential function includes only the first two terms of the power series expansion

$$e^x = 1 + x + \frac{x^2}{2!} + \cdots$$

in the limit as x approaches zero. The error associated with each of these asymptotic expansions decreases in the limit as $x \to 0$. For $x < 0.1$ radians, the asymptotic approximation for $\sin x$ is better than 2 parts in 10,000, while the approximation for e^x is better than 5 parts in 1,000. As we allow x to approach closer to zero the approximations improve: for $x < 0.01$, the approximation for e^x is better than 5 parts in 100,000, while the approximation for $\sin x$ is better than 2 parts in 10^7.

With the exception that an asymptotic series is not unique, the properties of asymptotic series are quite similar to those of power series:

Theorem 3.6 *Not every function has an asymptotic expansion (for example, e^x).*

Theorem 3.7 *Every infinitely differentiable function has an asymptotic expansion.*

Theorem 3.8 *An asymptotic expansion is not unique. It is possible for different functions to have the same asymptotic expansion or approximation.*

Theorem 3.9 *If a function has an asymptotic power series expansion, it is the only expansion.*

Theorem 3.10 *Asymptotic series can be added, subtracted, multiplied, exponentiated, and divided (as long as there is no division by zero, unless zero in the denominator is canceled by a zero in the numerator). The resulting series will be another asymptotic series.*

Theorem 3.11 *The asymptotic expansion of a function can be integrated term by term. The result will be the asymptotic expansion of the integral of the function.*

Theorem 3.12 *Term-by-term differentiation of an asymptotic expansion may not be valid. This is because small oscillating terms within the asymptotic expansion can become large when differentiated.*

3.7 Examples

This section presents a number of examples to demonstrate how the techniques in this chapter are applied in practical problems.

3.7.1 Definite Integrals

In analyzing applied problems, it is not uncommon to encounter expressions or results that cannot be evaluated in terms of tabulated or known functions. In cases where the result or expression is a definite integral, it may be possible to approximate or evaluate the value of the integral using a series expansion, even if the anti-derivative, or the solution to the indefinite integral, cannot be found.

For example, consider the integral

$$\int_0^1 e^{-x^2} dx.$$

This integral cannot be solved using traditional techniques, as the anti-derivative of the function is not known in terms of elementary functions. However, it is possible to express the integrand as a power series and then integrate term by term. To evaluate the integral

$$\int_0^1 e^{-x^2} dx,$$

we first find the series expansion for the integrand by substituting $-x^2$ for x in the Maclaurin series of e^x to obtain

$$e^{-x^2} = 1 - x^2 + \frac{x^4}{2!} - \frac{x^6}{3!} + \cdots + \frac{(-x^2)^n}{n!} + \cdots$$

This series expansion is then substituted for the integrand to obtain

$$\int_0^1 e^{-x^2} dx = \int_0^1 \left(1 - x^2 + \frac{x^4}{2!} - \frac{x^6}{3!} + \cdots + \frac{(-x^2)^n}{n!} + \cdots\right) dx.$$

Integrating term by term results in the expression

$$\int_0^1 e^{-x^2} = \left[x - \frac{x^3}{3} + \frac{x^5}{5.2!} - \frac{x^7}{7.3!} + \cdots + \frac{(-1)^n x^{2n+1}}{(2n+1)n!)} + \cdots\right]_0^1.$$

Evaluating at the limits of integration, we find that

$$\int_0^1 e^{-x^2} = 1 - \frac{1}{3} + \frac{1}{5.2!} - \frac{1}{7.3!} + \cdots = \sum_{n=0}^{\infty} \frac{(-1)^n}{(2n+1)n!}.$$

This expression can be evaluated to determine both an approximate value for the integral and the error associated with the approximation. For example, say we need to know the numerical value of this integral with an uncertainty of less than 0.01. We first recognize that the given series is an alternating series, so the error in considering only the first few terms of the series will be less than the absolute value of the first neglected term. If the series is summed up to and including the nth term, then the first neglected term will be the $(n+1)$th term. So it is the $(n+1)$th term that needs to be less than the stipulated error, i.e.,

$$\frac{1}{(2(n+1)+1)(n+1)!} = \frac{1}{(2n+3)(n+1)!} < 0.01.$$

For $n = 3$, we find that

$$\frac{1}{(2n+3)(n+1)!} = 0.004,$$

so that the finite summation

$$\sum_{n=0}^{3} \frac{(-1)^n}{(2n+1)n!}$$

gives an approximate value for the integral within the stipulated error.

3.7.2 Asymptotic Approximations

While asymptotic series are useful for approximating the value of a function $f(x)$, such approximations cannot be made arbitrarily accurate for any value of $x \neq 0$. This is because asymptotic expansions involve partial sums, not infinite summations, and the approximation therefore cannot be made arbitrarily close.

In Section 3.7, we stated that in general, the error associated with an asymptotic approximation is smaller in absolute value and has the same sign as the first neglected term of the partial sum [8]. In many cases, allowing the variable to approach a limiting value will reduce the error of an asymptotic approximation. However, if the number of terms included in the asymptotic expansion is increased, the approximation may get worse. **Stirling's approximation** provides an excellent example of this type of behavior. Stirling's approximation, or **Stirling's formula,** is

$$\ln n! \approx \left(n + \frac{1}{2}\right)\ln n - n + \ln\sqrt{2\pi}. \tag{3.20}$$

This approximation is actually the first two terms of the asymptotic series expansion

$$\ln n! \approx \left(n + \frac{1}{2}\right)\ln n - n + \ln\sqrt{2\pi} + \frac{1}{12n} - \frac{1}{360n^2} + \cdots$$

If we fix the number of terms in the partial sum to the first two terms and allow the value of the variable n to increase, we will find that for $n > 10$, the asymptotic approximation

$$\ln n! \approx \left(n + \frac{1}{2}\right)\ln n - n + \ln\sqrt{2\pi}$$

is accurate to within 6 parts in 10,000. If, on the other hand, we were to fix the value of n and include additional terms from the asymptotic expansion in the partial sum, the approximation to $\ln(n!)$ would get worse.

3.7.3 Series Approximations of Functions

3.7.3.1 Rule of 72

In finance, there are rules of thumb for estimating the doubling time for an investment. These are referred to as the **rule of 72**, the **rule of 70,** and the **rule of 69.3,** and they are useful in performing mental calculations [11]. When the interest on an investment is compounded periodically, the number of periods needed for the initial investment to double in value is estimated by taking the **rule number** (e.g., 72), and dividing it by the interest rate:

$$\text{Number of periods} = 75 \div \%InterestRate. \tag{3.21}$$

For example, if an investment has an interest rate of 9% compounded annually, then the number of years it would take for the investment to double would be

$$\text{Number of years} = 72 \div 9 = 8 \text{ years.}$$

These rules can be applied to exponential (or geometric) growth. Note that the rules are equally applicable for estimating the halving time of a decay process. The choice of the rule number is a matter of the application or preference. The choice of 72 is convenient for most commonly encountered interest rates, as the number 72 has many small divisors: 1, 2, 3, 4, 6, 8, 9, and 12. The choice 69 is more accurate approximation in general and for continuously compounding interest [11].

It is "interesting" (pun intended) to note that these approximations are derived from the power series expansion of the natural logarithm. If an initial principle investment of P is invested at a compound interest rate of $r\%$ (per period), then the value V of the investment after t periods is given by

$$V = P(1 + \frac{r}{100})^t.$$

To find the number of periods it takes for the investment to double in value, we set $V = 2P$, to get

$$V = 2P = P(1 + \frac{r}{100})^t.$$

To solve for the number of investment periods t, we take the natural logarithm of both sides of the equation to get

$$\ln 2 = t \ln(1 + \frac{r}{100}).$$

Rearranging this equation, we find that t, the number of periods required for the investment to double in value is

$$t = \frac{\ln 2}{\ln(1 + \frac{r}{100})}.$$

While this expression can be evaluated using a scientific calculator, the expression can be simplified further so that the calculation can be mentally performed. Recall that the power series expansion for $\ln(1 + x)$ is

$$\ln(1 + x) = x - \frac{x^2}{2} + \frac{x^3}{2} - \dots$$

So our expression for t can be expanded to read:

$$t = \frac{\ln 2}{(r/100) - \frac{(r/100)^2}{2} + \frac{(r/100)^3}{3} - \dots}.$$

For small values of r, the higher-order terms are powers of $(r/100)^n$ and are negligible. It is therefore possible to truncate the series expansion and consider only the first term of the series, with the doubling time t given as approximately

$$t \cong \frac{\ln 2}{(r/100)}.$$

Note that $\ln 2$ is approximately 0.693, so that

$$t \cong \frac{69.3}{r},$$

which is the rule of 69.3. For smaller values of r, it is possible to use a further simplification,

$$t \cong \frac{72}{r},$$

which has the advantage of being an easy mental calculation.

3.7.3.2 Russian Roulette

As mentioned in the previous example, the rule of thumb applies equally well for determining the halving time of a decay process. As a (morose) example of a decay process, consider the game Russian roulette. If there is a single bullet in a revolver cylinder with c chambers, then the probability of being shot on the first pull of the trigger is $1/c$. The probability of surviving the first pull is then $1 - 1/c$. If we continue playing for N cycles, spinning the cylinder before each pull of the trigger, the probability of surviving N pulls of the trigger is $(1 - 1/c)^N$. Therefore, the probability of being shot before N pulls of the trigger is

$$P = 1 - \left(1 - \frac{1}{c}\right)^N. \tag{3.22}$$

So what is the expected number of cycles N such that half the people pursuing this game would be shot? Setting $P = 1/2$, we have

$$\frac{1}{2} = 1 - \left(1 - \frac{1}{c}\right)^N.$$

Taking the natural logarithm of both sides of the equation and solving for N, we find that

$$N = \frac{-\ln 2}{\ln\left(1 - \frac{1}{c}\right)}.$$

Using the series expansion for $\ln(1 + x)$ to expand the denominator, we obtain

$$N = \frac{-\ln 2}{\frac{-1}{c} - \frac{\left(\frac{-1}{c}\right)^2}{2} - \frac{\left(\frac{-1}{c}\right)^3}{3} + \cdots},$$

For $c > 1$, we can get an approximate answer by neglecting every term of the series except for the first, to find that

$$N \cong \frac{-\ln 2}{\left(\frac{-1}{c}\right)} \cong \frac{0.693}{\left(\frac{1}{c}\right)} = c(0.693).$$

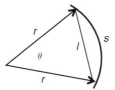

Figure 3.2 A length of arc verse a chord length.

For a revolver with 6 chambers ($c = 6$), this works out to be approximately 4 pulls of the trigger.

3.7.3.3 Railroad Surveying

In railroad and road surveying, long radius curves are laid out as a series of chords rather than as a circular arc. The curve is laid out by staking the ends of individual chords on the circumference of the curve. The length of a chord depends on the "degree of curvature" of the arc, and it is chosen so that the difference between the length of the circular arc and the length of the straight chord is less than the surveying uncertainty [12].

If r is the radius of the curve and θ the intercepted angle, then the length of arc s is given by

$$s = r\theta,$$

while the length of the associated chord l is given by

$$l = 2r \sin \frac{\theta}{2}.$$

In order to correctly lay out the curve, the difference between the length of the arc and the chord length must be less than the surveying tolerance. To estimate this difference, we can expand the expression for the chord length l as a power series:

$$l = 2r \sin \frac{\theta}{2} = 2r \left(\frac{\theta}{2} - \frac{\theta^3}{8 \cdot 3!} + \frac{\theta^5}{32 \cdot 5!} + \cdots \right).$$

Note that the series expansion is an alternating series, so that the error introduced by considering only the first few terms of the series is less than the first neglected term. Using only the first two terms of the series, the difference between the arc length and the chord length is approximately

$$s - l \cong r\theta - 2r \left(\frac{\theta}{2} - \frac{\theta^3}{3!} \right) = \frac{\theta^3}{24},$$

with an error less than the third term,

$$\frac{2r\theta^5}{32 \cdot 5!} = \frac{r\theta^5}{1,920}.$$

3.7.3.4 Parabolic Telescope Mirrors

Parabolic mirrors have the useful property that they change incoming plane waves into converging (reflected) spherical waves, which come to a focus. This is exactly the type of behavior you want from a telescope mirror: take an incoming plane wave from a distant object such as a star and produce an image (converging spherical wave) of the star.

Unfortunately, the grinding process used to form optical surfaces does not naturally produce a parabolic surface. Rather, the grinding process naturally produces a spherical surface, and spherical mirrors do not produce "nice" focused images. Instead, they produce images that suffer from spherical aberrations. In practice, parabolic telescope mirrors are produced by first grinding out an "appropriate" spherical surface. The outer regions of this sphere are then preferentially polished out to approximate the desired parabola.

What is the "appropriate" spherical surface? Well, it is one that requires a minimal amount of polishing and removal of material to produce a parabola with the desired focal length f. A series expansion is used to determine the radius R of the initial spherical surface and the amount of material that must be removed in order to produce a parabola of focal length f [13].

If we describe the cross section of the desired telescope mirror as a parabola with its vertex located at the origin, then the cross section of the sphere that approximates a parabola is shown in Figure 3.3.

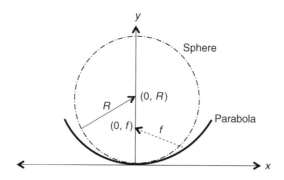

Figure. 3.3 Parabola and sphere with vertices at the origin.

The equation for a parabola with its vertex at the origin is

$$y_p = \frac{x^2}{4f},$$ (3.23)

while the equation for the spherical cross section whose center of curvature is located on the y-axis at $y_s = R$ is

$$(y_s - R)^2 + x^2 = R^2.$$

Solving for y_s, we find that

$$y_s = R - \sqrt{R^2 - x^2} = R\left(1 - \sqrt{1 - \left(\frac{x}{R}\right)^2}\right).$$

When the quantity $|x| / R < 1$ (as is the case for a telescope mirror), the square root term found on the right-hand side of the equation can be expanded using the following binomial series:

$$1 - \sqrt{1 - \left(\frac{x}{R}\right)^2} = 1 - \frac{1}{2}\left(\frac{x}{R}\right) - \frac{1}{8}\left(\frac{x}{R}\right)^4 - \cdots$$

Substituting this series expansion into the expression for y_s gives

$$y_s = \frac{x^2}{2R} + \frac{x^4}{8R^3} + \cdots$$

This is the power series representation of the cross section of the sphere. Note that the first term of the series expansion,

$$\frac{x^2}{2R},$$

can be interpreted as the equation for a parabola if we let

$$4f = 2R.$$

In other words, the initial sphere that best approximates a parabolic mirror with focal length f is one whose radius is twice the desired focal length:

$$R = 2f.$$

While the first term of our series expansion represents the desired parabola, the higher-order terms of the series represent the difference between the parabola and the approximating sphere. These higher-order terms describe the amount of material that needs to be removed from the sphere in order to create the parabola. Setting $R = 2f$ and subtracting the equation for the parabola from the series expansion for the sphere, we find that the difference between the sphere and the parabola is given by

$$\Delta y = y_s - y_p = \frac{1}{64}\frac{x^4}{f^3} + \cdots$$

This series represents the amount of material that needs to be removed from the edge of the mirror. For a typical amateur telescope mirror with a diameter of 200 mm and a focal length of 1,000 mm, the thickness of the material that must be removed is about

$$\Delta y \simeq \frac{1}{64}\frac{100^4}{1000^3} \simeq 0.0016 \text{ mm.}$$

This thickness corresponds to roughly one wavelength of infrared radiation.

3.7.4 Indeterminate Forms

L'Hôpital's rule is frequently used to evaluate indeterminate expressions of the form ∞ / ∞ or $0/0$. This method requires taking the derivative of the functions in both the numerator and denominator:

$$\lim_{x \to a}\frac{f(x)}{g(x)} = \lim_{x \to a}\frac{f'(x)}{g'(x)}. \qquad (3.24)$$

In situations where the derivatives of the functions are complicated or problematic (for example, $\arctan x$ or $x!$), it may be better to evaluate the limit using series expansions. As a simple example, consider finding the limit of $(\sin x) / x$ as $x \to 0$. Although this expression has the indeterminate form $0/0$, rather than use L'Hôpital's rule, let us replace $\sin x$ with its infinite series expansion. Taking the limit of the resulting infinite series expression gives

$$\lim_{x \to 0}\frac{\sin x}{x} = \lim_{x \to 0}\frac{\left(x - \frac{x^3}{3!} + \frac{x^5}{5!} - \frac{x^7}{7!} + \cdots\right)}{x} = \lim_{x \to 0}\left(1 - \frac{x^2}{3!} + \frac{x^4}{5!} - \frac{x^6}{7!} + \cdots\right) = 1.$$

The series method for evaluating indeterminate forms is most useful for limits where the indeterminate form is $0/0$ and $x \to 0$. To understand why this should be the case, note that a Maclaurin series is an expansion in powers of x around the origin ($x = 0$). In the limit as x approaches 0, a Maclaurin series simply reduces to a constant term. This is exactly the behavior demonstrated in the preceding example.

3.7.5 Summing Numerical Series

In some situations, we may need to sum an unfamiliar numerical series. In some of these cases, it may be possible to find the sum of the unfamiliar series in a mathematical handbook. Series involving the natural numbers and or the powers of natural numbers, such as

$$1^2 + 2^2 + 3^2 + \cdots + n^2 = \frac{n}{6}(n+1)(n+2),$$

occur frequently, and their sums can typically be found in mathematical tables.

In cases where we cannot find a series in a mathematical reference, it may be possible to sum the series using the techniques developed in this chapter. It may be that the numerical series is simply the series expansion of a function, evaluated for a particular value of x. For example, consider having to sum the series

$$1 - \frac{1}{3} + \frac{1}{5} - \frac{1}{7} + \frac{1}{9} - \cdots$$

At first glance, the task of summing this unfamiliar series might appear a bit daunting. However, searching through a table of series expansions, you might notice the similarity between this series and the series expansion

$$\arctan x = x - \frac{x^3}{3} + \frac{x^5}{5} - \frac{x^7}{7} + \frac{x^9}{9} - \cdots$$

Setting $x = 1$ in the series expansion for arctan x, we get the expression

$$\arctan 1 = 1 - \frac{1}{3} + \frac{1}{5} - \frac{1}{7} + \frac{1}{9} - \cdots$$

Having identified the relationship between the unfamiliar series and the series expansion for arctan x, we can simply evaluate arctan x at $x = 1$ and set it equal to the sum of the unfamiliar series:

$$\arctan 1 = \frac{\pi}{4} = 1 - \frac{1}{3} + \frac{1}{5} - \frac{1}{7} + \frac{1}{9} - \cdots$$

In this example, we were able to analytically sum a series of numerical values because it was recognizable as the series expansion of arctan x (evaluated at $x = 1$). The following examples demonstrate how such situations may arise in applied problems.

Figure 3.4 One-dimensional crystal.

3.7.6 The Madelung Constant

In solid-state physics, the **Madelung constant** (α) is used to determine the electrostatic potential of an ion located in a crystal lattice [14]. The electrostatic potential energy of the ion can be expressed as

$$U(r) = \alpha \frac{ke^2}{r},$$

where α is the Madelung constant for the crystal lattice. Determining the Madelung constant involves a series summation referred to as the **lattice sum**. As a simple example, consider a one-dimensional chain of alternating negative and positive ions, each separated by the distance r.

Each ion has two immediate neighbors of the opposite charge at a distance of r. The next two nearest ions are of like charge but at a distance of $2r$, the third nearest ions are of opposite charge but at a distance of $3r$, and so on. The electrostatic potential energy of an ion located in the one-dimensional lattice is the sum of the electrostatic potential energy due to the presence of all the other ions, i.e.,

$$U(r) = \frac{ke^2}{r} + \frac{ke^2}{r} - \frac{ke^2}{2r} - \frac{ke^2}{2r} + \frac{ke^2}{3r} + \frac{ke^2}{3r} - \frac{ke^2}{4r} - \frac{ke^2}{4r} + \cdots$$

Combining like terms, we find that the electrostatic potential takes the form of the series expression

$$U(r) = 2\frac{ke^2}{r}\left(1 - \frac{1}{2} + \frac{1}{3} - \frac{1}{4} + \cdots\right).$$

We are now left in the position of having to determine the sum of the series

$$\left(1 - \frac{1}{2} + \frac{1}{3} - \frac{1}{4} + \cdots\right).$$

We might be tempted to simply approximate the sum of this alternating harmonic series by summing the first n terms of the series and noting that the approximation is good to within the absolute value of the first neglected term $(n + 1)$.

However, a quick look at a reference book or table of series expansions would reveal the similarity between this sum and the Maclaurin series

$$\ln(x+1) = x - \frac{x^2}{2} + \frac{x^3}{3} - \frac{x^4}{4} + \cdots$$

Setting $x = 1$ in the Maclaurin series expansion, we find that

$$\ln(1+1) = \ln 2 = 1 - \frac{1}{2} + \frac{1}{3} - \frac{1}{4} + \cdots$$

In this case, we can analytically solve the summation and determine that the potential energy of an ion lattice is given by

$$U(r) = (2\ln 2)\frac{ke^2}{r},$$

where the factor 2(ln2) is the Madelung constant α for the one-dimensional lattice.

3.7.7 Numerical Computations

The series representation of a function is useful in numerical computations when we need an approximate numerical value for the function. For example, if we need the approximate value of $\ln(1.1)$, we can set $x = 0.1$ in the series representation for $\ln(x+1)$,

$$\ln(x+1) = x - \frac{x^2}{2} + \frac{x^3}{3} - \frac{x^4}{4} + \cdots,$$

to find that

$$\ln(1.1) = 0.1 - \frac{(0.1)^2}{2} + \frac{(0.1)^3}{3} - \frac{(0.1)^4}{4} + \cdots$$

Summing the first four terms of the series representation and rounding off to the fifth decimal place, we find that

$$\ln(1.1) \simeq 0.1 - 0.005 + 0.000333 - 0.000025 + \cdots \simeq 0.9531.$$

As the series representation is a convergent alternating series, we know that the absolute value of the error in this approximation is less than the first neglected term of the series, which is

$$\left| \frac{(0.1)^5}{5} \right| = 0.000002.$$

4

Complex Infinite Series

4.1 Complex Numbers

Complex analysis is common in engineering and the physical sciences, as it frequently simplifies calculations or affords deeper insight into the nature of a problem. Many physical theories have been generalized by allowing the physical variable to be complex rather than real. For example, when the index of refraction n of a material is taken to be complex $(n - ik)$, the theory incorporates refraction as well as the absorption of light. The introduction of complex notation can simplify the analysis of many physical systems – for example, oscillatory behavior (electrical and/or mechanical) naturally lends itself to complex analysis.

The use of complex power series is important for defining functions of complex variables. With a theory of complex functions, it is possible to attach meaning to functions such as

$$\cos i, e^i, \text{ and } i^{-2i}$$

and to solve equations that have no real solutions, such as

$$\cos z = 2.$$

Believe it or not, such expressions do occur in applied problems. Functions of complex variables are useful for solving the differential equations that occur in fields as diverse as quantum physics and fluid mechanics.

Recall that for a **quadratic equation**

$$az^2 + bz + c = 0,$$

the general solution for the unknown z is given by the **quadratic formula**

$$z = \frac{-b \pm \sqrt{b^2 - 4ac}}{2a}.$$

The quantity under the square root symbol,

$$b^2 - 4ac,$$

is called the **discriminant**. In situations where the discriminant is negative, we find ourselves having to take the square root of a negative number. However, only nonnegative numbers have real roots; for example, the square root of four,

$$\sqrt{4},$$

has two roots, 2 and −2. Now consider the square root of negative four:

$$\sqrt{-4}.$$

Neither 2^2 or $(-2)^2$ will return a result of −4. In this example, there are no real solutions. There are, however, imaginary solutions. It is possible to find a solution if we introduce a new type of number, called an **imaginary number**. The notation or "unit" used to denote an imaginary number is

$$i = \sqrt{-1},$$

where it is understood that

$$i^2 = -1.$$

If we adopt the use of i, then the square roots of −4 can be expressed as

$$\sqrt{-4} = \sqrt{(-1)4} = \sqrt{-1}\sqrt{4} = i\sqrt{4},$$

which is an imaginary number. There are actually two such roots, both of which are imaginary:

$$\sqrt{-4} = i\sqrt{4} = \pm 2i.$$

These roots can be verified by simple multiplication:

$$2i \cdot 2i = 4i^2 = -4 \quad \text{and} \quad -2i \cdot -2i = 4i^2 = -4.$$

Numbers such as

$$-i, \quad i\sqrt{2}, \quad \text{and} \quad \pm 2i$$

are referred to as **pure imaginary numbers**, while the numbers

$$i^2 = -1, -2, \text{ and } 3$$

are **real**.

A **complex number** is a number with both a real and an imaginary component. Such numbers can be denoted as $x + iy$, or the sum of two terms, one of which is real and the other imaginary. For example, the following are complex numbers:

$$3 + i, \ 1 - 5i, \text{ and } 1 + 2i.$$

Consider the complex number

$$1 + 2i.$$

The **imaginary part** of the complex number $1 + 2i$ is the coefficient of the i term, or the imaginary term, which is 2. The **real part** of the complex number $1 + 2i$ is the coefficient of the real term, which is 1. It is important to understand that the real and imaginary coefficients of a complex number are *both real numbers*. It might seem odd that the real and imaginary parts of a complex number are both real. However, we can make sense of this by thinking of i as a bookkeeping device that allows us to keep track of real and imaginary numbers separately. In most computations, i acts as a placeholder or label, with the understanding that when it is squared, i^2 should be replaced by -1. As we will see in the following sections, the fact that the real and imaginary parts of a complex number are both real numbers will allow us to interpret complex numbers geometrically, either as a point in a complex plane or as a vector.

4.1.1 Geometrical Representations of Complex Numbers

It is possible to represent the complex number $z = 1 + 2i$ as a pair of real numbers $z = (1, 2)$. This notation is suggestive of a point in an xy-plot. The real and imaginary components of a complex number can be interpreted as coordinates in the xy-plane. This type of plot is called an **Argand diagram** and is shown in Figure 4.1.

When the xy-plane is used to display complex numbers, it is referred to as the **complex plane**. If we were to plot a pure imaginary number $0 + iy$ on the complex plane, the point would lie on the y-axis. Hence, the y-axis is referred to as the **imaginary axis,** and the imaginary part of a complex number is designated as the y-coordinate:

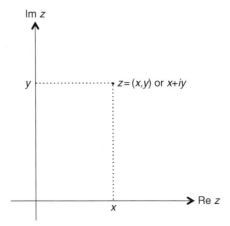

Figure 4.1 The Argand diagram.

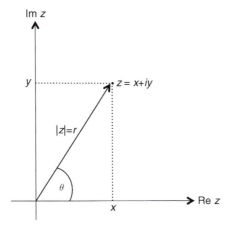

Figure 4.2 A vector in the complex plan.

$$\text{Im } z = y. \tag{4.1}$$

Similarly, if we were to plot a real number $x + 0i$ on the complex plane, the point would lie on the x-axis. Hence, the x-axis is referred to as the **real axis,** and the real part of a complex number z is represented by the x-coordinate:

$$\text{Re } z = x. \tag{4.2}$$

The Argand diagram also suggests that we would be justified in interpreting a complex number as a two-dimensional vector in the complex plane; see Figure 4.2.

The **length** r of the vector is given by

$$|z| = r = \sqrt{x^2 + y^2}$$

and is also referred to as the **modulus** or **absolute value** of z:

$$\text{mod } z = |z|. \qquad (4.3)$$

The angle θ that the vector makes with the positive x-axis is referred to as the **angle** of z and also as the **phase** or the **argument** of z:

$$\theta = \arg z = \text{angle of} z. \qquad (4.4)$$

Given a complex number $z = x + iy$, it is possible to determine the argument of z using the inverse trigonometric formula

$$\theta = \arctan (y/x). \qquad (4.5)$$

The $x + iy$ representation of a complex number is referred to as the **rectangular form** of the complex number, as x and y each represent the rectangular coordinates of a point in the complex plane. Interpreting a complex number as a vector allows us to use vector manipulations to define the algebra of complex numbers.

In addition to the rectangular form $x + iy$, there is another form for expressing a complex number, called the **polar form**. In polar form, a complex number is written as

$$z = r(\cos \theta + i \sin \theta). \qquad (4.6)$$

The polar form of a complex number has many advantages and is useful for defining elementary functions of complex numbers. The **elementary functions** are exponential functions, logarithmic functions, trigonometric functions, and their inverses, as well as powers and roots of complex numbers. It is possible to locate a point using not only the rectangular coordinates (x, y) but also the real polar coordinates (r, θ).

From Figure 4.3, we can see that we can use the relations

$$x = r \cos \theta,$$
$$y = \sin \theta,$$

and

$$r = \sqrt{x^2 + y^2}$$

to transform between polar and rectangular coordinates. With these relationships in mind, it is simple to transform the rectangular form of a complex number to the polar form:

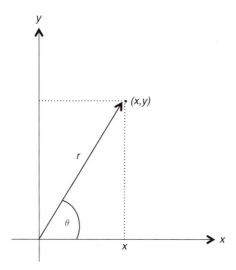

Figure 4.3 Locating a complex number with polar coordinates.

$$z = x + iy = r\cos\theta + ir\sin\theta = r(\cos\theta + i\sin\theta).$$

The polar form for a complex number can be further simplified using **Euler's equation,**

$$e^{i\theta} = \cos\theta + i\sin\theta, \tag{4.7}$$

to obtain

$$z = x + iy = r(\cos\theta + i\sin\theta) = e^{i\theta}, \tag{4.8}$$

where r is the **length, modulus,** or **absolute value** of z and is always taken to be positive, while θ is referred to as the **angle, argument,** or **phase** of z and can take on either positive or negative values. The advantage of polar notation is that calculation involving complex numbers is typically easier to perform using the polar form rather than the rectangular form $x + iy$.

As an example, consider the complex number $1 + i$. This number can be represented in a number of different ways. In rectangular coordinates, it is the point $(1, 1)$, while in polar coordinates it is the point

$$\left(\sqrt{2}, \frac{\pi}{4}\right),$$

and it can even be written as

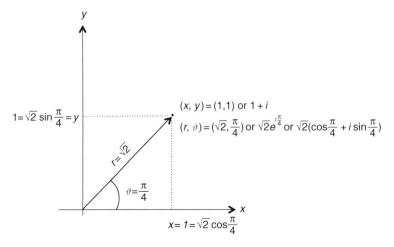

Figure 4.4 The different ways to express a complex number.

$$1 + i = \sqrt{2}\left(\cos\frac{\pi}{4} + i\sin\frac{\pi}{4}\right) = \sqrt{2}e^{i\frac{\pi}{4}}.$$

These various ways of expressing a complex number are illustrated in Figure 4.4.

It is important to note that the angle is given in **radians**, not **degrees**. Many of the mathematical formulas that you are familiar with are only correct when the angle is measured in radians. For example, the small angle approximations

$$\tan\theta \cong \sin\theta \cong \theta$$

are only correct when θ is given in radians. Degrees are convenient to work with but only when explicitly working with functions of angles or intentionally seeking an angle. For example, the expression

$$\sin 30° = \frac{1}{2}$$

is perfectly acceptable, as it is clear that we are only considering angles. If, on the other hand, we are working through a numerical calculation that requires us to find the arcsin of 1/2, the correct answer is

$$\arcsin\frac{1}{2} = 0.5235\ldots$$

and not

$$\arcsin\frac{1}{2} = 30°.$$

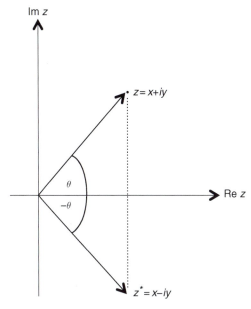

Figure 4.5 A complex number and its conjugate.

Expressing the solution as an angle would only be correct in situations where we were explicitly seeking an angle.

Before moving on to the algebra of complex numbers, we need to introduce the concept of the complex conjugate of z. **The complex conjugate** of a complex number z, represented as z^* or $\bar{z}$, is obtained by changing the sign of the imaginary component of the complex number. For example, if the complex number is

$$z = (x + iy),$$

then its complex conjugate is

$$z^* = (x - iy). \tag{4.9}$$

Notice that only the sign of the i term has changed. When plotted on the complex plane, a complex number and its complex conjugate are mirror images of one another, reflected across the x-axis (the real axis); see Figure 4.5.

In polar form, the complex conjugate of $z = r(\cos\theta - i\sin\theta)$ is written as

$$z^* = r(\cos\theta - i\sin\theta) = r(\cos(-\theta) + i\sin(-\theta)) = re^{-i\theta}.$$

In polar form, we find the complex conjugate by simply changing the sign of i in the exponent:

$$z = re^{i\theta} \text{ and } z^* = re^{-i\theta}.$$

Notice that, when written in polar form, a complex number and its conjugate have the same positive value for r, but their phases θ are oppositely signed. The fact that complex numbers appear in **conjugate pairs**,

$$re^{i\theta} \text{ and } re^{-i\theta},$$

is useful when finding roots of polynomials.

4.1.2 Complex Addition and Subtraction

Addition and subtraction of complex numbers are straightforward. The real and imaginary components of a complex number are added (subtracted) separately using the ordinary rules of algebra, i.e.,

$$z_1 \pm z_2 = (x_1 + iy_1) \pm (x_2 + iy_2) = (x_1 \pm x_2) + i(y_1 \pm y_2).$$

Why complex numbers should be added (or subtracted) in this way becomes apparent when we think of a complex number as a vector with real and imaginary components. If we were adding pure imaginary numbers, no number of additions (or subtractions) would cause the sum to become real and leave the y-axis (the imaginary axis). Similarly, when adding (or subtracting) real numbers, no number of additions (or subtractions) would cause the number to become imaginary and leave the x-axis (the real axis).

A quirky way to think of complex numbers is to think of them as a fruit basket containing two different kinds of fruit, say, apples and oranges. While apples and oranges are both fruits, i.e.,

$$\#\text{fruits} = (\#\text{apples} + \#\text{oranges}),$$

no number of apples can be added to make an orange, and similarly, no number of oranges can become an apple, so that when combining fruit baskets (complex numbers), the numbers of apple apples and oranges are added up separately.

4.1.3 Complex Multiplication

To find the product of two complex numbers z_1 and z_2, simply multiply the numbers out in full, with the understanding that $i^2 = -1$. For example,

$$z_1 z_2 = (x_1 + iy_1)(x_2 + iy_2) = (x_1 x_2 - y_1 y_2) + i(x_1 y_2 + y_1 x_2).$$

As an example, multiplying $(1 - 2i)$ by $(3 + 4i)$, we find that

$$(1 - 2i)(3 + 4i) = 3 + 4i - 6i + 8 = (11 - 2i).$$

4.1.4 Complex Division

When written in rectangular form, division by a complex number,

$$\frac{z_1}{z_2} = \frac{(x_1 + iy_1)}{(x_2 + iy_2)},$$

is not an effortless task. Division of this kind requires the use of complex conjugation in order to be able to express the results in rectangular form. The "trick" is to multiply both the numerator and the denominator by the complex conjugate of the denominator:

$$\frac{z_1}{z_2} = \frac{(x_1 + iy_1)}{(x_2 + iy_2)} = \frac{(x_1 + iy_1)(x_2 - iy_2)}{(x_2 + iy_2)(x_2 - iy_2)} = \frac{(x_1 + iy_1)(x_2 - iy_2)}{x_2^2 + y_2^2}.$$

Notice that when the denominator is multiplied by its complex conjugate, the denominator becomes real. The formal name for this process is **rationalizing the denominator**. Completing the multiplication in the numerator, we find that

$$\frac{z_1}{z_2} = \frac{(x_1 x_2 - y_1 y_2)}{x_2^2 + y_2^2} + i\frac{(x_1 x_2 + y_1 y_2)}{x_2^2 + y_2^2},$$

and the quotient is in rectangular form. As an example, consider dividing $3 + 4i$ by $1 - 2i$:

$$\frac{(3 + 4i)}{(1 - 2i)} = \frac{(3 + 4i)(1 + 2i)}{(1 - 2i)(1 + 2i)} = \frac{3 + 6i + 4i - 8}{1 + 4} = \frac{-5 + 10i}{5} = -1 + 2i.$$

It is important to recognize that the division and multiplication of complex numbers is sometimes simpler if the complex numbers are written in the polar form

$$re^{i\theta}.$$

In this case, multiplication takes the form

$$z_1 z_2 = r_1 e^{i\theta_1} r_2 e^{i\theta_2} = r_1 r_2 e^{i\theta_1} e^{i\theta_2} = r_1 r_2 e^{i(\theta_1 + \theta_2)},$$

while division looks like

$$\frac{z_1}{z_2} = \frac{r_1 e^{i\theta_1}}{r_2 e^{i\theta_2}} = \frac{r_1}{r_2} e^{i(\theta_1 - \theta_2)}.$$

As an example, we will use division by a complex number to prove the useful identity $1/i = -i$. The complex number

$$\frac{1}{i}$$

can be put in the form $(x + iy)$ by rationalizing by the denominator. Multiplying the numerator and denominator by the complex conjugate of the denominator, we find that

$$\frac{1}{i} = \frac{1}{i}\left(\frac{i^*}{i^*}\right) = \frac{1}{i}\left(\frac{-i}{-i}\right) = \frac{-i}{1} = -i.$$

While this is an odd-looking identity, it is frequently useful and worth committing to memory.

4.1.5 Finding Mod z or $|z|$

Complex conjugation is not only useful in division; it can also be used to determine the modulus or absolute value of a complex number. Recall that the modulus or absolute value of a complex number z is given by

$$\text{mod } z = |z| = r = \sqrt{x^2 + y^2},$$

which is a positive square root, as both x and y are real coefficients. Multiplying z by its complex conjugate z^*, we get

$$zz^* = (x + iy)(x - iy) = x^2 + y^2 = r^2,$$

where r is always a real number that is greater than or equal to zero. This will also be the case if we work in polar coordinates:

$$zz^* = \left(re^{i\theta}\right)\left(re^{-i\theta}\right) = r^2.$$

So in both polar and rectangular coordinates, we find that

$$zz^* = r^2 = |z|^2 = (\text{mod } z)^2.$$

Because the modulus is a real number that is greater than or equal to zero, in taking the square root of this expression, we find that

$$\text{mod } z = |z| = \sqrt{zz^*}. \tag{4.10}$$

The fact that the square of the modulus of a complex number is equal to the complex number times its conjugate, i.e.,

$$|z|^2 = zz^*,$$

is worth memorizing, as this identity is frequently used when manipulating complex expressions. Two additional identities that are frequently useful are

$$|z_1 z_2| = |z_1||z_2|$$

and

$$\left|\frac{z_1}{z_2}\right| = \frac{|z_1|}{|z_2|}.$$

These properties of the modulus follow directly from the identities for the complex conjugate of a complex expression, which is discussed in the following section.

4.1.6 The Complex Conjugate of an Expression

Another labor-saving device that is worth memorizing is *the complex conjugate of an operation on complex numbers is equal to the operation on the conjugates*. The following simple examples illustrate this property. The complex conjugate of a sum or difference of two complex numbers is equal to the sum (or difference) of the complex conjugates:

$$(z_1 \pm z_2)^* = z_1^* \pm z_2^*.$$

Similarly, the complex conjugate of the product or quotient of two complex numbers is the product or quotient of the complex conjugates:

$$(z_1 z_2)^* = z_1^* z_2^* \text{ and } \left(\frac{z_1}{z_2}\right)^* = \frac{z_1^*}{z_2^*}.$$

These properties are useful for manipulating complex expressions, as it may be difficult or labor-intensive to reduce a complex expression to the rectangular form $x + iy$ in order to find the complex conjugate. The division of two complex numbers given in rectangular form is a good example. Consider the complex number

$$z = \frac{3 + 4i}{1 - 2i}.$$

To find the complex conjugate, simply change the signs of all the i terms to get

$$z^* = \left(\frac{(3 + 4i)}{(1 - 2i)}\right)^* = \frac{3 - 4i}{1 + 2i}.$$

This method for finding the complex conjugate is more expedient than attempting to rearrange the quotient z into the form $x + iy$ by first rationalizing the denominator, as we did in Section 4.1.4.

It is important to be careful when using this method to find a complex conjugate, as an expression may contain "buried" or "hidden" is'. For example, while the complex conjugate of

$$z = (x + iy)$$

is

$$z^* = (x - iy),$$

as both x and y are real coefficients, the complex conjugate of

$$z = z_1 + iz_2$$

is

$$z^* = z_1{}^* - iz_2{}^*,$$

not

$$z = z_1 - iz_2,$$

because both z_1 and z_2 are themselves complex. A few more complex conjugate expressions worth memorizing are

$$(z^*)^* = z,$$
$$z + z^* = 2 \operatorname{Re} z = 2x,$$
$$z - z^* = 2i \operatorname{Im} z = 2y,$$

and

$$\frac{z}{z^*} = \left(\frac{x^2 - y^2}{x^2 + y^2}\right) + i\left(\frac{2xy}{x^2 + y^2}\right).$$

4.1.7 Complex Equations

Recall that a complex number can be represented as a pair of real numbers: $z = (x, y)$. This suggests that two complex numbers are **equal** only when they correspond to the same point in the complex plane. That is, their real parts must be equal and their imaginary parts must be equal. For example, the equation

$$x + iy = 1 + 2i$$

is true if and only if

$$x = 1 \text{ and } y = 2.$$

That is, a complex equation is actually two "real" equations, one for each part or component of the complex number. Consider the complex equation

$$(z)^2 = (x + iy)^2 = i$$

Squaring the left-hand side of this equation, we find that

$$(x^2 - y^2) + i(2xy) = i.$$

This equation can be rewritten to make the distinction between the real and imaginary components more explicit:

$$(x^2 - y^2) + i(2xy) = (0 + i).$$

This expression can be thought of as two separate equations, one for each coefficient of the complex number. Equating like coefficients, we get the two equations

$$x^2 - y^2 = 0$$

and

$$2xy = 1.$$

The first equation, $x^2 - y^2 = 0$, determines the real component of our complex solution. Rewriting this equation, we find that

$$y^2 = x^2,$$

so that

$$y = x.$$

Substituting this result into the second equation, $2xy = 1$, gives the result

$$2x^2 = 1$$

and

$$-2x^2 = 1.$$

Recall that x is known to be a real number, so x^2 cannot be negative. This leaves us with a single equation,

$$2x^2 = 1.$$

So

$$x^2 = \frac{1}{2},$$

and we thus have two solutions:

$$x = y = \frac{1}{\sqrt{2}} \text{ and } x = y = -\frac{1}{\sqrt{2}}.$$

Therefore, the solution to the complex equation

$$(z)^2 = (x + iy)^2 = i$$

when given in rectangular form is

$$z = (x + iy) = \pm\left(\frac{1}{\sqrt{2}} + \frac{1}{\sqrt{2}}i\right).$$

Although we have solved the equation

$$(x + iy)^2 = i$$

in rectangular coordinates, it is also possible to solve the equation in polar coordinates. Solving the equation using the polar form $re^{i\theta}$ is revealing, as it affords us a geometric interpretation of complex multiplication and thus ties together the concepts of a modulus and an argument. To gain this perspective, we need to rewrite the complex equation in polar form.

First consider the right-hand side of the equation

$$(x + iy)^2 = i.$$

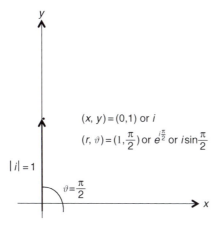

Figure 4.6 Equivalent expressions for i.

When the number $z = i = 0 + i$ is plotted on the complex plane (x, y), it falls on the y-axis (the imaginary axis) at the point $(0, 1)$; see Figure 4.6. The imaginary number i can be represented by an **associated vector** directed from the origin to the point $(0, 1)$, where the length of the vector is given by

$$r = \sqrt{x^2 + y^2} = \sqrt{0^2 + 1^2} = 1.$$

In polar form, $re^{i\theta}$, the number i is written as

$$i = e^{i\frac{\pi}{2}}.$$

One way to obtain this expression is by inspecting Figure 4.6. The argument or phase of i is the angle of inclination θ of the vector as measured counterclockwise from the positive x-axis (the real axis), which in this case is clearly $\pi/2$ radians:

$$\theta = \arg i = \frac{\pi}{2}.$$

Furthermore, r, the length, modulus, or absolute value of the complex number $z = i$, is given by

$$r = \sqrt{0^2 + 1^2} = \mathrm{mod}\ i = |i| = \sqrt{ii^*} = 1.$$

This leads us to conclude that when written in polar form $re^{i\theta}$, the number $z = i$ is

$$i = (1)e^{i\frac{\pi}{2}} = e^{i\frac{\pi}{2}}.$$

Now that we have managed to transform the right-hand side of the equation into polar form, let us turn our attention to the left-hand side of the equation:

$$(x + iy)^2 = i.$$

The left-hand side has the unknown variable $z = x + iy$. In a rectangular coordinate system, we would be seeking the real parameters (x, y) that solve the equation. In a polar coordinate system, we are seeking the real parameters (r, θ) that provide a solution. Therefore, we can simply rewrite the unknown on the left-hand side as

$$(x + iy) = re^{i\vartheta}.$$

Thus, in polar coordinates, the original complex equation

$$(x + iy)^2 = i$$

appears as

$$(re^{i\vartheta})^2 = e^{i\frac{\pi}{2}}.$$

A moment of consideration will reveal that the equation is much simpler to solve in polar form than in the original rectangular form. Taking the square root of both sides of the equation, we find two solutions:

$$z = re^{i\vartheta} = \left(e^{i\frac{\pi}{2}}\right)^{\frac{1}{2}} = \pm\left(e^{i\frac{\pi}{4}}\right).$$

Therefore,

$$re^{i\vartheta} = \pm\left(e^{i\frac{\pi}{4}}\right),$$

and equating like terms in this expression leads us to conclude that the values for the real parameters (r, θ) are

$$r = 1, \quad \theta = \frac{\pi}{4}.$$

It is illuminating to consider our two solutions from a "vector perspective." Figure 4.7 illustrates that the positive solution

$$z = e^{i\frac{\pi}{4}}$$

is associated with a vector of length $|z| = r = 1$, which subtends an angle of $\pi/4$ radians with respect to the positive real axis.

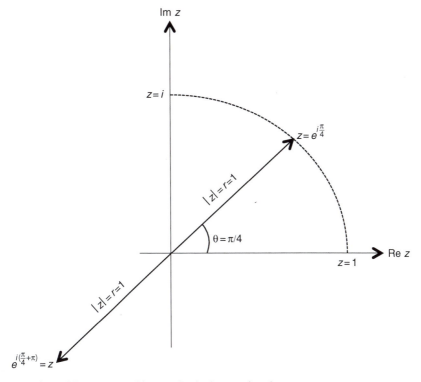

Figure 4.7 A vector and its negative in the complex plane.

Recall that the parameter r is **taken to be positive.** If this is the case, then how do we interpret the negative solution

$$z = -\left(e^{i\frac{\pi}{4}}\right)?$$

The negative solution is associated with a vector of length $r = 1$ but points in the opposite direction to the positive solution, i.e., $\pi/4 + \pi$. To see this, consider where -1 lies in the complex plane and how it is expressed using complex notation:

$$-1 = 1e^{i\pi}.$$

Therefore, the negative solution can be expressed as

$$z = -e^{i\frac{\pi}{4}} = e^{i\pi}e^{i\frac{\pi}{4}} = e^{i\left(\frac{\pi}{4}+\pi\right)},$$

which is a vector of length $r = 1$.

These two points can be rewritten in the form $(x + iy)$ to demonstrate that they are identical to the points found by solving the equation in rectangular coordinates. Recall from Section 4.1.1 that the rectangular and polar forms are related by the equation

$$z = (x + iy) = r(\cos \theta + i \sin \theta) = re^{i\theta}.$$

Thus, the positive solution corresponds to the point

$$e^{i\frac{\pi}{4}} = \cos \frac{\pi}{4} + i \sin \frac{\pi}{4} = \left(\frac{1}{\sqrt{2}} + \frac{1}{\sqrt{2}}i\right),$$

and the negative solution corresponds to the point

$$e^{i(\frac{\pi}{4}+\pi)} = e^{i\frac{5\pi}{4}} = \cos \frac{5\pi}{4} + i \sin \frac{5\pi}{4} = \left(-\frac{1}{\sqrt{2}} - \frac{1}{\sqrt{2}}i\right).$$

This confirms that the points found by solving the equation in polar coordinates are identical to those we obtained when we first solved the equation in rectangular coordinates.

We are now in a position to interpret complex multiplication (division) geometrically. When written in polar form, the product of two complex numbers z_1 and z_2 is

$$z_1 z_2 = r_1(\cos \theta_1 + i \sin \theta_1)r_2(\cos \theta_2 + i \sin \theta_2).$$

Completing the multiplication, we find that

$$z_1 z_2 = r_1 r_2 \left((\cos \theta_1 \cos \theta_2 - \sin \theta_1 \sin \theta_2) + i(\sin \theta_1 \cos \theta_2 + \cos \theta_1 \sin \theta_2)\right).$$

Use of the angle addition formulas for cosine and sine simplifies this expression to

$$z_1 z_2 = r_1 r_2(\cos (\theta_1 + \theta_2) + i \sin (\theta_1 + \theta_2)).$$

From this formula, we can conclude that *in complex multiplication, the moduli of complex numbers are multiplied and their arguments are summed*, i.e.,

$$z_1 z_2 z_3 \ldots = (r_1 r_2 r_3 \ldots)e^{i(\theta_1 + \theta_2 + \theta_3 + \ldots)}.$$

Geometrically, a complex product can be associated with a vector whose length is equal to the product of the lengths and whose argument (phase) or angle is the sum of the angles (see Figure 4.8).

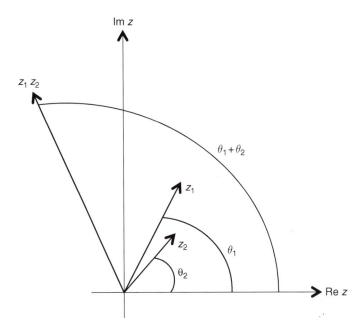

Figure 4.8 Multiplication by a complex number is equivalent to a rotation.

Following a similar line of thought, we can conclude that *complex division amounts to dividing moduli and subtracting arguments*, i.e.,

$$\frac{z_1}{z_2} = \left(\frac{r_1}{r_2}\right) e^{-i(\theta_1 - \theta_2)}.$$

With this new geometrical picture in mind, you may wish to go back and review Sections 4.1.3 and 4.1.4, where the expressions for complex multiplication and division were originally introduced.

The geometric picture of complex multiplication (Figure 4.8) is interesting in that it allows us to interpret multiplication by i as a $\pi/2$ counterclockwise rotation and multiplication by i^2 as a rotation π of radians. To fully understand this interpretation, consider multiplying the complex number $x + iy$ by i:

$$i(x + iy) = -y + ix.$$

Observe that the coefficient x, which originally designated the real component of the complex number, now designates the imaginary component, while the coefficient y, which initially designated the imaginary component, now designates the real component. This exchange of "identities" amounts to a counterclockwise rotation through $\pi/2$ radians (see Figure 4.9). It follows that

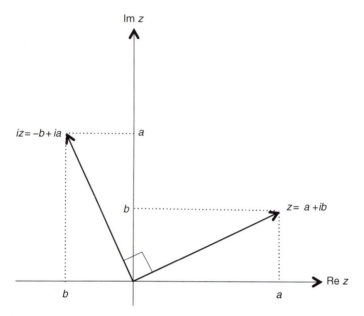

Figure 4.9 Multiplication by i is equivalent to a 90-degree rotation.

multiplication by $i^2 = -1$ is equivalent to a counterclockwise rotation by π radians. That is, two 90-degree rotations ($i \times i$) leave the vector pointing in the opposite direction:

$$i^2 z = i^2(x + iy) = -(x + iy) = -x - iy.$$

The rules for complex multiplication, that moduli multiply and arguments add, lead us to a very important result, namely, that powers of a complex number z are given by the equation

$$z^n = r^n(\cos n\theta + i \sin n\theta) = r^n e^{in\theta}. \tag{4.11}$$

The method for determining the **complex roots** z_n of an equation such as this one is important in constructing solutions to linear ordinary differential equations and to partial differential equations that describe physical systems.

We are now at a point where we have naively stumbled into the field of complex analysis. The "mathematical pool" we were wading in has unexpectedly become very deep, and the pool now contains objects such as $\cos i$, $\ln i$, and $\cos z = 2$. We are going to momentarily step back from this "deep end" of the pool and will return to these topics later in this chapter. There is, however, one special case that we are already prepared to consider, the case where $r = 1$. When $r = 1$, the powers of a complex number are given by the equation

$$z^n = (\cos n\theta + i \sin n\theta) = e^{in\theta}.$$

When rewritten, this equation is recognizable as DeMoivre's theorem:

$$e^{in\theta} = (e^{i\theta})^n = (\cos \theta + i \sin \theta)^n = (\cos n\theta + i \sin n\theta).$$

This equation is useful for deriving trigonometric identities for functions of multiple angles, i.e., $\cos n\theta$ and $\sin n\theta$. For example, the double angle formula for both sine and cosine can be found simultaneously because a complex equation is equivalent to two real equations. Applying DeMoivre's theorem, we find that

$$\cos 2\theta + i \sin 2\theta = (\cos \theta + i \sin \theta)^2.$$

Completing the multiplication and collecting like terms yields a result in rectangular form $x + iy$:

$$\cos 2\theta + i \sin 2\theta = (\cos^2 \theta - \sin^2 \theta) + 2i(\sin \theta \, \cos \theta).$$

Equating like coefficients gives us two real equations, one for each component of the complex number:

$$\cos 2\theta = (\cos^2 \theta - \sin^2 \theta),$$
$$\sin 2\theta = 2 \sin \theta \cos \theta.$$

These results are the double angle formulas found in mathematical handbooks.

4.1.8 Powers and Roots

We have intentionally neglected some important details concerning complex numbers for the sake of telling the story. One of these points is that, unlike real numbers, *complex numbers have no natural ordering*. In others words, it is meaningless to ask, for example, whether $1 + 2i$ is greater or less than $1 - 2i$. Complex numbers do not fall on a number line; rather, they fall into or "paint" a region of the complex plane. This aspect of complex numbers will be significant when we consider the convergence properties of a complex series representation of a complex function.

Secondly, *complex numbers are not uniquely determined*. While the absolute value ($|z|$), modulus (mod z), and length (r) of a complex number are uniquely determined, the argument, the phase, or the angle θ is not. For example, when plotted on the complex plane, the numbers

$$\sqrt{2}e^{i\frac{\pi}{4}} = \sqrt{2}\left(\cos \frac{\pi}{4} + i \sin \frac{\pi}{4}\right)$$

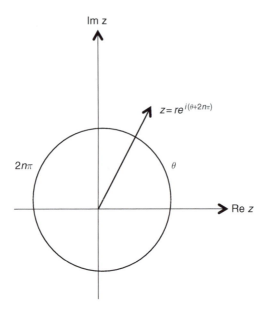

Figure 4.10 The argument or phase of a complex number is multivalued.

and

$$\sqrt{2}e^{i(\frac{\pi}{4}+2\pi)} = \sqrt{2}\left(\cos\left(\frac{\pi}{4}+2\pi\right) + i\sin\left(\frac{\pi}{4}+2\pi\right)\right)$$

plot at the same location, namely, $1 + i$. A rotation of 2π in the complex plane brings a vector back around to its original position. As a matter of fact, for every integer n, the argument θ of a complex number z could be replaced by $\theta + 2\pi n$, and we would obtain the same z. That is, ***arg z is multivalued***, and every rotation of 2π radians plots at the same location on the complex plane (see Figure 4.10).

Recall from Section 4.1.5 that the power of a complex number expressed in polar form is given by the expression

$$z^n = \cos n\theta + i\sin n\theta = r^n e^{in\theta}.$$

This number could itself be written in polar form as

$$z^n = r^n e^{in\theta} = \rho e^{i\phi} = w.$$

Notice that this equation is true if and only if

$$r^n = \rho \text{ and } n\theta = \phi.$$

To understand why this is the case, recall that two complex numbers are equal,

$$a + ib = c + id,$$

if and only if $a = c$ and $b = d$. While this property is expressed in rectangular form (x, y), the statement is equally true in polar form (r, θ), so that in order for the equation

$$r^n e^{in\theta} = \rho e^{i\phi}$$

to be true, the conditions

$$r^n = \rho \text{ and } n\theta = \phi$$

must also be true. For two complex numbers to be equal, they must have the same absolute value or length:

$$|z^n| = r^n = \rho = |w|.$$

This implies that r is the nth root of ρ:

$$r = \sqrt[n]{\rho} = \rho^{\frac{1}{n}}.$$

It is important to remember that the length r (and ρ) of a complex number is always taken to be positive, so we only need consider the positive nth roots of ρ and can neglect the negative roots. For example, if we were seeking a solution to the equation

$$\rho = \sqrt{4},$$

we would most likely think of 2, not -2.

Now let us turn our attention to the arguments or angles of our two complex numbers. Just as the absolute values and the lengths must be equal, so must the arguments:

$$n\theta = \phi.$$

There is, however, a subtlety here. Recall that the argument of a complex number is multivalued and that for every integer n, the argument ϕ of a complex number w can be replaced by $\phi + 2\pi n$. So if the two arguments are equal, then the equation

$$n\theta = \phi + 2\pi n$$

must also be true, and we get the result

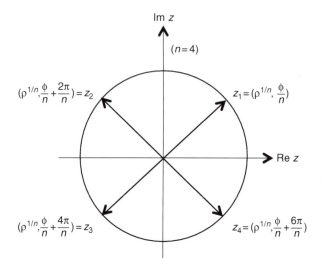

Figure 4.11 The fourth roots of a complex number.

$$\theta = \frac{\phi}{n} + \frac{2\pi n}{n}.$$

Bringing everything together, we can say that the complex number

$$w = \rho e^{i\phi} = (re^{i\theta})^n = z^n$$

has n distinct roots (z_n), given by the formula

$$\sqrt[n]{w} = z_n = \rho^{\frac{1}{n}} e^{i\left(\frac{\phi}{n} + 2\pi\frac{m}{n}\right)} = \rho^{\frac{1}{n}}\left(\cos\left(\frac{\phi}{n} + \frac{2\pi m}{n}\right) + i\sin\left(\frac{\phi}{n} + \frac{2\pi m}{n}\right)\right). \quad (4.12)$$

The integers $m = 0, 1, 2, 3, \ldots, n-1$ will produce distinct roots, and all other values of m simply reproduce one of these roots. If we were to plot the nth roots in the complex plane, we would find n roots of equal length, uniformly separated by $2\pi/n$ radians, with the initial root located at the point $(\rho^{1/n}, \phi/n)$; see Figure 4.11.

It is interesting to observe that the sum of the n nth roots of a complex number will always be zero. This can be seen in Figure 4.11 by noting that the x and y components of all the roots cancel when summed.

As an example, consider finding the cube roots of 8:

$$z = \sqrt[3]{8}.$$

The number 2 is clearly a cube root of 8, but it is not the only cube root; there are two more. The number 2 is the cube root of 8 when we are only considering

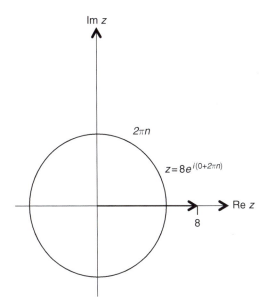

Figure 4.12 Location the real number 8 in the complex plane.

positive roots and neglecting complex roots. In order to identify all of the nth roots of a number, we first express the number in polar form. In our case, we would write 8 as

$$8 = 8e^{i(0+2\pi)}.$$

On the complex plane, the number 8 plots at the location (8, 0). This is a vector from the origin to the point (8, 0) with a length or modulus of 8, and it is parallel to the x-, or real, axis. Lying parallel to the x-axis, the vector has an included angle of zero radians; therefore, the argument is zero. However, recall that the arguments of a complex number are multivalued, and every rotation of $2\pi n$ radians will leave the vector in the same orientation (Figure 4.12).

Having expressed 8 in polar form, we are now in a position to identify the cube roots:

$$z = \sqrt[3]{8} = (8e^{i2\pi n})^{\frac{1}{3}} = \sqrt[3]{8}e^{i\frac{2\pi n}{3}} = 2e^{i\frac{2\pi n}{3}}.$$

To find all roots, we simply index through the integer values of n to find

$$z = 2e^{0},\ 2e^{i2\pi/3},\ 2e^{i4\pi/3},\dots$$

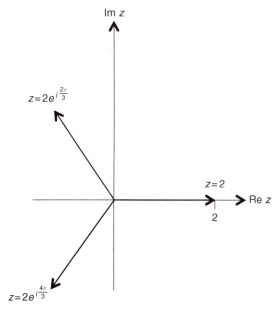

Figure 4.13 The cube roots of 8.

Only the integer values $n = 0, 1, 2$ give distinct roots, and all other integer values repeat one of these three roots. Figure 4.13 shows the three distinct cube roots plotted on the complex plane.

Note that the three cube roots of 8 sum to zero. This can be seen by examining Figure 4.13 and noting that the x and y components of all the roots sum to zero. In summary, to find the nth roots of a complex number z:

i *Express the complex number in polar form, $re^{i\theta}$.*
ii *Find the real positive nth root of r.*
iii *Divide the argument $(\theta + 2m\pi)$ by n.*
iv *The initial root is given by $(r^{1/n}, \theta/n)$.*
v *The remaining n^{th} roots are found by repeatedly adding $2\pi/n$ to the argument.*

4.2 Complex Infinite Series

It is relatively simple to take the convergence tests developed for infinite series of real terms and extend those tests for infinite series of complex terms. The key to extending the convergence tests developed in Chapter 2 to the complex plane is recognizing that a complex number can be represented as a pair of real numbers. If the complex number is given in the form $x + iy$, the pair of real numbers is (x, y); if given in polar form $re^{i\theta}$, the pair of real numbers is (r, θ).

Consider, for example, an infinite series of complex numbers

$$S = z_1 + z_2 + z_3 + \ldots$$

Expanding the infinite series in rectangular form, we find that

$$S = (x_1 + iy_1) + (x_2 + iy_2) + (x_3 + iy_3) + \ldots$$

Collecting like terms, we obtain the expression

$$S = (x_1 + x_2 + x_3 + \ldots) + i(y_1 + y_2 + y_3 + \ldots),$$

which is clearly an expression for a complex infinite series. Notice, however, that the series

$$x_1 + x_2 + x_3 + \ldots$$

and

$$y_1 + y_2 + y_3 + \ldots$$

are both infinite series of real numbers. This means that the tests developed in Chapter 2 for analyzing infinite series of real terms can be applied to each of these series separately. Furthermore, the convergence of a complex series is defined just as it was in Section 2.2 but with the added stipulation that in order for a complex series to converge, both of the series $x_1 + x_2 + x_3 + \ldots$ and $y_1 + y_2 + y_3 + \ldots$ must converge. For example, consider the nth partial sum of an infinite series of complex terms:

$$S_n = (x_1 + x_2 + \ldots + x_n) + i(y_1 + y_2 + \ldots + y_n).$$

If the partial sum S_n approaches the limit S as $n \to \infty$, then the series is said to be convergent and S is taken as the sum of the series:

$$\lim_{n \to \infty} S_n \to S = X + iY.$$

In order for S_n to approach a limit, both the real and the imaginary parts of the series must converge, i.e.,

$$\lim_{n \to \infty} (x_1 + x_2 + \ldots + x_n) \to X$$

and

$$\lim_{n \to \infty} (y_1 + y_2 + \ldots + y_n) \to Y.$$

We may now suspect that many of the concepts and tests developed for a series of real terms can simply be extended to accommodate a series whose terms are complex. For example, the concept of absolute convergence can be readily applied to a complex series. For an infinite series of complex terms, the absolute value of each complex term is given by the expression

$$|z| = r = \sqrt{x^2 + y^2} = \text{mod } z = \sqrt{zz*},$$

which is always positive. That is, a series of absolute-valued terms is a positive term series, and any convergence test developed for positive terms series (Section 2.3.2) can be used to test a complex series for absolute convergence. For example, consider testing the complex series

$$\sum_{n=0}^{\infty} z^n = \sum_{n=0}^{\infty} (re^{i\theta})^n$$

for convergence. Series of this form occur when wave disturbances (mechanical or electromagnetic) are summed using the complex expression $re^{i\theta}$ for an oscillation. This approach to summation is far simpler than attempting to sum a number of waves represented by sine functions; see the example in Section 4.6.6. The series

$$\sum_{n=0}^{\infty} (re^{i\theta})^n$$

is clearly a **complex geometric series** with the common ratio $z = re^{i\theta}$. It is very simple to test this series for convergence using the ratio test. The ratio of two consecutive terms of the geometric series is

$$\rho_n = \frac{z^{n+1}}{z^n} = \frac{(re^{i\theta})^{n+1}}{(re^{i\theta})^n} = re^{i\theta}.$$

Recall that, by the ratio test, a series is absolutely convergent when

$$\rho = \lim_{n \to \infty} |\rho_n| < 1,$$

so that our geometric series is convergent when

$$\rho = \lim_{n \to \infty} |\rho_n| = \lim_{n \to \infty} |re^{i\theta}| = \lim_{n \to \infty} \sqrt{re^{i\theta}re^{-i\theta}} = r < 1.$$

Therefore, the series

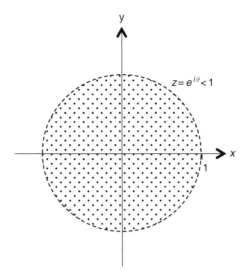

Figure 4.14 The disk of convergence.

$$\sum_{n=0}^{\infty} (re^{i\theta})^n$$

converges if and only if $r < 1$.

A subtle new feature must be pointed out. Recall from Section 4.1.6 that a complex number is not uniquely determined. While the absolute value $|z|$ or length r of a complex number is uniquely determined, the argument or phase angle θ is not. This means that our complex geometric series is convergent for all complex numbers with an absolute value or length less than 1, *independent of their argument*. If all such numbers were plotted on the complex plane, they would fill or "paint" the interior of a disk as illustrated in Figure 4.14; this disk is referred to as **the disk of convergence**.

The disk of convergence is a generalization of the concept of the interval of convergence. In fact, an interval of convergence, say $(-a, a)$, is the interval along the x-axis (the real axis) that falls within the disk of convergence.

The complex geometric series we considered in the last example is simply a special case of a **complex power series**

$$\sum a_n z^n, \tag{4.13}$$

where $a_n = 1$. In general, both the terms z and a_n can be complex numbers. Such complex power series have many of the same useful features as real power series, if not more. In fact, these types of power series turn up naturally as solutions to ordinary and partial differential equations.

The calculus of complex functions of a complex variable is so encompassing that our "usual" calculus (functions of a real variable) can be viewed as a special case of complex calculus in which the complex number is taken to be $z = x + i0$. In fact, it will be useful for us to adopt this perspective for the rest of this chapter. In Chapter 3, we focused on understanding and manipulating real power series of the form

$$\sum a_n x^n.$$

Such a power series can be regarded as a special case of a complex power series, where $z = x + i0$, i.e., $y = 0$. A similar observation could be made for the imaginary part of a complex series; recall that the imaginary part of a complex number is actually a real coefficient. You can probably surmise that the convergence of a complex power series is determined by the convergence properties of two real series (the real and the imaginary parts). This implies that the convergence tests developed for real series are equally applicable to complex power series.

We state without proof that *the power series theorems given in Section 3.2 are also true for complex power series.* This extension or continuation from real to complex variables is done with the understanding that the interval of convergence is to be replaced by a disk of convergence. For example, we have the following theorems:

Theorem 4.1 *If a function has a power series expansion or representation, it is unique.*

Theorem 4.2 *Convergent series may be added, subtracted, or multiplied, and the resulting series is convergent at least within the disk of convergence common to all series.*

Theorem 4.3 *Convergent power series may be divided as long as there is no division by zero, unless a zero occurring in the denominator series is canceled by a zero occurring in the numerator series. The quotient series converges around the origin inside the smallest of the following three disks: the disk of convergence of the numerator, the disk of convergence of the denominator, and the disk defined by the shortest distance to the origin (in the complex plane) where the denominator is zero.*

Theorem 4.4 *A power series may be substituted into another power series provided that the interval of convergence of the substitute series lies within the disk of convergence of the original series.*

Theorem 4.5 *A power series may be differentiated or integrated term by term, and the resulting series will converge to the derivative or integral of the function represented by the original series within the disk of convergence of the original series.*

Note that Theorem 4.3 as been heavily modified compared to how it originally appeared in Section 3.2. This modification limits our consideration to expressions that are *analytic*. Informally speaking, a function is said to be analytic in a region when it is infinitely differentiable at every point in that region and single-valued. Note that if a function were not differentiable to all orders (i.e., analytic) at a point, then it could not be expanded in a Taylor series around that point, as a Taylor series expansion requires consecutive differentiation.

To summarize, a complex power series converges inside a circle extending up to the nearest point (on the complex plane) for which the function is no longer analytic. We will revisit this behavior in greater detail in Sections 4.4 and 4.5.

4.3 Determining the Disk of Convergence

As with real power series, complex power series are useful because the series expansions can be treated as if they were polynomials, so that within the region of convergence, series expansions can be added, subtracted, multiplied, divided, differentiated, and integrated.

While real power series converge within some interval on the x-axis (the real axis), complex power series converge within a disk centered at a point on the complex plan, the disk of convergence. We can demonstrate this by using the ratio test. Recall that we used the ratio test in Section 3.2 to demonstrate that power series converge on an interval centered around the origin.

Consider the generalized complex power series

$$f(z) = \sum_{n=0}^{\infty} a_n (z - z_0)^n,$$

where a_n is a complex coefficient, z_0 is a complex constant, and z is a complex variable. Application of the ratio test revels that the complex series is absolutely convergent for

$$\rho = \lim_{n \to \infty} \left| \frac{a_{n+1} (z - z_0)^{n+1}}{a_n (z - z_0)^n} \right| < 1. \qquad (4.14)$$

Taking the limit of the quotient, we find that the general solution is given by

$$|z - z_0| < \frac{a_n}{a_{n+1}}.$$

There is a simple geometric interpretation for this expression. This interpretation can be made apparent by expressing the constant a_n / a_{n+1} as r, which results in the equation

$$|z - z_0| < r.$$

This formula describes a disk of convergence centered at the point z_0 on the complex plane. The constant r is (if you haven't already guessed) referred to as the **radius of convergence.** Equations of the form

$$|z - z_0| < r,$$

occur frequently, so it is worth memorizing that such an equation describes a disk in the complex plane. To recognize that the equation

$$|z - z_0| < r$$

describes the **interior of a circle**, rewrite the complex variable z and the complex constant z_0 in rectangular form, so that

$$z = x + iy \text{ and } z_0 = a + ib.$$

Substituting these expressions for z and z_0 in the equation for the disk, we find that

$$|z - z_0| = |(x + iy) - (a + ib)| = |(x - a) + i(y - b)| < r.$$

Recall that the absolute value or length of a complex number is writing as $|z| = \sqrt{x^2 + y^2}$, so that

$$|(x - a) + i(y - b)| = \sqrt{(x - a)^2 + (y - b)^2} < r.$$

Squaring both sides of this expression, we get

$$(x - a)^2 + (y - b)^2 < r^2.$$

This inequality describes the interior of a circle of radius r centered at the point $(x, y) = (a, b)$; see Figure 4.15.

In other words, a complex series converges for values of z near the center (a, b) of a disk and diverges when z is too far away. The radius of convergence r specifies how far one can stray from the center of the disk and still expect the

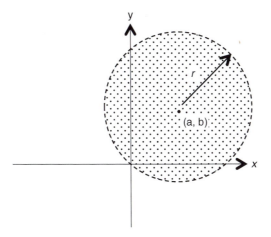

Figure 4.15 The disk of convergence.

series to converge. If a series converges for all values of z, then the radius of convergence is infinite. Note that we have not addressed the convergence properties of a series on the boundary

$$|z - z_0| = r.$$

The reason for this is that the behavior of a series on a boundary can be very complicated. Recall that for $\rho = 1$, the ratio test is inconclusive; the series may be convergent or divergent. This means that it is possible for a series to be convergent at some locations on the boundary and divergent at others. We will leave such issues for more advanced texts.

As a concrete example, consider determining the radius of convergence for the complex power series

$$\sum_{n=0}^{\infty} \frac{(z + 1 + i)^n}{3^n}.$$

Applying the ratio test, we find that

$$\rho = \lim_{n \to \infty} \left| \frac{3^n (z + 1 + i)^{n+1}}{3^{n+1} (z + 1 + i)^n} \right| < 1$$

so that

$$\rho = \lim_{n \to 0} \left| \frac{z + 1 + i}{3} \right| = \left| \frac{z + 1 + i}{3} \right|.$$

Thus, the series converges for

$$|z + (1 + i)| < 3.$$

Rewriting the equation so that it appears in the form $|z - z_0|$, we find that

$$|z - (1 - i)| < 3.$$

This describes a disk with radius $r < 3$ centered at $z_0 = -1 - i$, or the point $(-1,-1)$.

4.4 Functions of Complex Variables

Elementary functions are functions of a single variable involving combinations of: powers, roots, and exponential, logarithmic, trigonometric, and inverse trigonometric functions. It is possible to generalize the elementary functions so that they are defined for both real and complex variables. When extended to incorporate complex variables, a number of the elementary functions turn out to have properties that make them extremely useful in mathematics and the physical sciences. These functions are known as **analytic functions,** and when combined with a calculus for complex variables, they provide a powerful analysis tool that is widely used in the physical sciences.

The definitions of the elementary functions can be extended to complex variables through the use of complex power series expansions. The extension of the exponential function e^x from the real number line onto the rest of the complex plane (e^z) is a cornerstone of complex analysis in the physical sciences. We would like to incorporate complex variables into our definitions of the elementary functions in such a way that the functions reduce to familiar, real expressions when $z = x + i0$, i.e., z is a real number. It is possible to define e^z in this fashion by simply substituting $z = x + iy$ for x in the Taylor series expansion of the function e^x, giving

$$e^z = \sum_{n=0}^{\infty} \frac{z^n}{n!}. \tag{4.15}$$

Use of the ratio test demonstrates that the preceding series representation converges for all z, as when we apply the ratio test, we find that

$$\rho(z) = \lim_{n \to 0} |\rho_n(z)| = \lim_{n \to 0} \left| \frac{n! z^{n+1}}{(n+1)! z^n} \right| = \lim_{n \to 0} \left| \frac{z}{n+1} \right| \to 0.$$

Therefore, we conclude that

$$\rho(z) < 1 \text{ for all } z,$$

and the series converges to the value e^z for all values of z.

It is useful to evaluate the series expansion for e^z along both the real and the imaginary axis. Along the real or x-axis, $y = 0$, so that the complex number $z = x + iy$ simple reduces to $z = x$, and the series expression reduces to the Taylor series for the real function e^x:

$$e^z = e^{(x+i0)} = \sum_{n=0}^{\infty} \frac{(x + i0)^n}{n!} = \sum_{n=0}^{\infty} \frac{x^n}{n!} = e^x.$$

While along the imaginary or y-axis, $x = 0$, so that the complex number $z = x + iy$ reduces to the expression $z = iy$ and the series expansion now looks like

$$e^z = e^{iy} = \sum_{n=0}^{\infty} \frac{(iy)^n}{n!}.$$

If we write out the series expansion, we get

$$e^{iy} = 1 + iy + \frac{(iy)^2}{2!} + \frac{(iy)^3}{3!} + \frac{(iy)^4}{4!} + \frac{(iy)^5}{5!} + \cdots$$

Completing the multiplications and collecting similar terms, we find that

$$e^{iy} = \left(1 - \frac{y^2}{2!} + \frac{y^4}{4!} - \cdots\right) + i\left(y - \frac{y^3}{3!} + \frac{y^5}{5!} - \cdots\right).$$

This is a revealing result when we recognize that the expressions contained within the sets of parentheses are the series expansions for $\cos y$ and $\sin y$, so that the expression simplifies to

$$e^{iy} = \cos y + i \sin y.$$

We have just derived the famous Euler formula!

Furthermore, although we do not provide a proof, the power series expansion

$$e^z = \sum_{n=0}^{\infty} \frac{z^n}{n!}$$

is the only definition of e^z that preserves the familiar formulas

$$e^{z_1} \cdot e^{z_1} = e^{z_1 + z_1} \text{ and } \left(\frac{d}{dx}\right)e^z = e^z.$$

With the complex series definition of e^z, it is now possible to use complex series to generalize the other elementary functions so that they are defined for both real or complex variables.

4.5 Laurent Series

Simply put, the Laurent series is a generalization or extension of the Taylor series expansion to the complex plane. In Section 3.3, we discussed how a function $f(x)$ of a real variable might be represented as the power series

$$f(x) = \sum_{n=0}^{\infty} a_n (x - c)^n.$$

In order to have such a power series expansion, the function must be differentiable to all orders at the point $x = c$, and the value of the remainder

$$R_n(x) = f(x) - \left[f(c) + (x - c)f'(c) + (x - c)^2 f''(c) + \cdots \right]$$

should approach zero. The function needs to be infinitely differentiable because the value of the power series coefficients a_n are determined from the terms of the Taylor series expansion, i.e.,

$$a_n = \frac{f^{(n)}(c)}{n!}.$$

Functions of a real or complex variable that are discontinuous will not have a power series representation around their "sharp corners," as differentiation is not defined at a discontinuity.

A function $f(z)$ of a complex variable that is infinitely differentiable (analytic) at $z = c$ can be represented by the power series

$$f(z) = \sum_{n=0}^{\infty} a_n (z - c)^n = \sum_{n=0}^{\infty} \frac{f^{(n)}(c)}{n!} (z - c)^n. \tag{4.16}$$

In this case, the unique power series representation will converge to the function inside a circle centered at the point $z = c$ and having a radius equal to the radius of convergence. If, however, there is at least one point inside the disk of convergence where the function is not analytic, we cannot expand the function in a Taylor series. This is analogous to the case of a real function not having a series expansion around a discontinuity. However, although a function of a complex variable cannot be expanded in a Taylor series around a point c where the function is not analytic (a "singularity"), the function can be expanded around a singularity using a Laurent series:

$$f(z) = \sum_{n=0}^{\infty} a_n (z - c)^n + \sum_{n=1}^{\infty} \frac{b_n}{(z - c)^n}. \qquad (4.17)$$

Singular points or **singularities** are points where a complex function is not analytic, i.e., the function is either not differentiable and or not single-valued. There are basically two types of singularities: branch points and isolated singular points. **Branch points** are locations where a complex function is multivalued; we first encountered multivalued functions in Section 4.1.6, when considering logarithms, complex powers, and complex roots. The different values of a multivalued function are referred to as different **branches** of the function. Multivalued functions can be treated as single-valued (i.e., as analytic) as long as we are careful to stay on a single branch. In short, branch points can be dealt with by first identifying the branch points of the function and then defining which branch we intend to stay on.

Isolated singular points, however, offer a new challenge that we have not yet considered. If a function $f(z)$ is analytic everywhere in a region except at the point $z = c$, then $z = c$ is called an **isolated singular point**. There are two types of isolated singular points: **poles** and **essential singularities**. If $f(z)$ is not finite at $z = c$ and there is an integer n such that the function

$$(z - c)^n f(z)$$

is analytic at $z = c$, then $f(z)$ has a **pole of order n** at $z = c$, where n is the smallest such integer. For example, consider the function

$$f(z) = \frac{1}{(z - 1)^2},$$

which has an isolated singularity at $z = 1$, as the function is not analytic (finite) at that point (the denominator is zero). In addition, the function

$$f(z) = (z - 1)^2 \frac{1}{(z - 1)^2}$$

is analytic at $z = 1$, and therefore the isolated singular point is a pole of order 2. Because the modified function

$$f(z) = (z - 1)^2 \frac{1}{(z - 1)^2}$$

is analytic at $z = 1$, this pole is referred to or classified as a **removable singularity**. If an isolated singularity is not a pole (i.e., is not removable), it is referred to as an **essential singularity**. For example, the function

$$f(z) = \sin\left(\frac{1}{z}\right)$$

has an essential singular point at $z = 0$.

Returning to the topic of Laurent series expansions, we note that a series expansion that contains both positive and negative powers of $z - z_0$,

$$f(z) = a_0 + a_1(z - z_0) + a_2(z - z_0)^2 + \cdots + \frac{b_1}{(z - z_0)} + \frac{b_2}{a_2(z - z_0)^2} + \cdots,$$

is referred to as a **Laurent series**. Note that the first summation of the Laurent series,

$$f(z) = \sum_{n=0}^{\infty} a_n(z - z_0)^n + \sum_{n=1}^{\infty} \frac{b_n}{(z - z_0)^n},$$

is a power series expansion in powers of $z - z_0$ around the point z_0, and if convergent, it would be convergent *inside* some disk of convergence around z_0.

The second summation in the Laurent series, i.e., the b term series, is a series of inverse powers of $z - z_0$ and is referred to as the **principle part** of the Laurent series. This portion of the Laurent series converges for

$$\left|\frac{1}{z - z_0}\right| < \text{ some constant } d.$$

This expression can be rewritten to make its interpretation more apparent:

$$|z - z_0| > 1/d.$$

In other words, series of this form would be convergent *outside* a disk of convergence centered on z_0. So in general, if a Laurent series is convergent, it converges in a region between two concentric circles. Recall from the convergence properties of powers series that it is possible for the inner circle to be a point and for the radius of the outer circle to extend infinitely; see Section 4.2.

As an example, consider the function

$$f(z) = \frac{1}{(z - 1)(z - 2)}.$$

This function is non-analytic at the points where the denominator is equal to zero, i.e., at the points $z = 1$ and $z = 2$. Both of these singularities are poles of order 1 and are removable. To find the Laurent series expansion for the function, we need to consider three separate regions: the region $|z| < 1$, the region $1 < |z| < 2$, and the region $|z| > 2$.

Using partial fraction decomposition, the function can be rewritten as

$$f(z) = \frac{1}{(z-1)(z-2)} = \frac{1}{z-2} - \frac{1}{z-1}.$$

In the region $|z| < 1$, we can express each term of the function as the sum of a geometric series:

$$\frac{1}{z-2} = -\frac{1}{2}\left(\frac{1}{1-\frac{z}{2}}\right) = -\frac{1}{2}\sum_{n=0}^{\infty}\left(\frac{z}{2}\right)^n$$

and

$$\frac{1}{z-1} = -\frac{1}{1-z} = -\sum_{n=0}^{\infty} z^n.$$

Combining the two series expressions results in the Laurent series

$$\frac{1}{(z-1)(z-2)} = \frac{1}{(z-2)} - \frac{1}{(z-1)} = \frac{1}{2}\sum_{n=0}^{\infty}\frac{z^n}{2^n} + \sum_{n=0}^{\infty}z^n$$

$$= \sum_{n=0}^{\infty}\left(-\frac{z^n}{2^{n+1}} + z^n\right) = \frac{1}{2} + \frac{3}{4}z + \frac{7}{8}z^2 + \cdots$$

In the region $1 < |z| < 2$, the geometric series expansion

$$\frac{1}{z-2} = -\frac{1}{2}\left(\frac{1}{1-\frac{z}{2}}\right) = -\frac{1}{2}\sum_{n=0}^{\infty}\left(\frac{z}{2}\right)^n$$

is still valid, as $|z/2| < 1$. However, the $1/(z-1)$ term needs to be expanded as

$$\frac{1}{z-1} = \frac{1}{z}\left(\frac{1}{1-\frac{1}{z}}\right) = \frac{1}{z}\sum_{n=0}^{\infty}\frac{1}{z^n} = \sum_{n=0}^{\infty}\frac{1}{z^{n+1}}$$

in order for the geometric series representation to be convergent. Once again, the two individual series expressions are combined to give the full Laurent series expansion,

$$\frac{1}{(z-1)(z-2)} = -\frac{1}{2}\sum_{n=0}^{\infty}\left(\frac{z}{2}\right)^n - \sum_{n=0}^{\infty}\frac{1}{z^{n+1}} = \cdots - \frac{1}{z^2} - \frac{1}{z} - \frac{1}{2} - \frac{z}{4} - \cdots$$

Finally, in the region $|z| > 2$, the $1/(z-2)$ term needs to be expanded as

$$\frac{1}{z-2} = \frac{1}{z}\left(\frac{1}{1-\frac{2}{z}}\right) = \frac{1}{z}\sum_{n=0}^{\infty}\left(\frac{2}{z}\right)^n = \sum_{n=0}^{\infty}\frac{2^n}{z^{n+1}},$$

in order for the geometric series expansion to be convergent. Therefore, in this final region, the Laurent series is given by the expression

$$\frac{1}{(z-1)(z-2)} = \sum_{n=0}^{\infty}\frac{2^n}{z^{n+1}} - \sum_{n=0}^{\infty}\frac{1}{z^{n+1}} = \sum_{n=0}^{\infty}\frac{2^n-1}{z^{n+1}} = \frac{1}{z^2} + \frac{3}{z^3} + \frac{7}{z^4} + \cdots$$

Three aspects of Laurent series should be pointed out. First, if a function $f(z)$ can be expanded around z_0 in a convergent Laurent series of the form

$$f(z) = a_0 + a_1(z-z_0) + a_2(z-z_0)^2 + \cdots + \frac{b_1}{(z-z_0)} + \frac{b_2}{a_2(z-z_0)^2} + \cdots,$$

where all the b_ns are zero, the function $f(z)$ is analytic at $z = z_0$, and z_0 is called a **regular point**. Second, if it happens that after some b_m (that is, not equal to zero), all the remaining b_ns are zero, then $f(z)$ is said to have a **pole of order m** at z_0; and in the special case where $m = 1$, $f(z)$ is said to have a **simple pole**. Third, if there are an infinite number of b_ns that are not equal to zero, then $f(z)$ is said to have an **essential singularity** at the point $z = z_0$. As we will discover in Chapter 5, we will need to be able to identify regular points and singular points of differential equations in order to find series expansion solutions for the differential equations at those points.

4.6 Examples

The following subsections present a number of examples of applied problems involving complex numbers.

4.6.1 Complex Index of Refraction

In optics, the index of refraction n is a real, dimensionless number specifying how fast light travels through a particular material. It is defined as the ratio

$$n = \frac{c}{v},$$

where c is the speed of light in vacuum and v is the **velocity** of light in the material. In air ($n_{air} = 1$), the **reflectance** (R) of a specific material is determined by its index of refraction using the equation

$$R = \left(\frac{n-1}{n+1}\right)^2.$$

In cases where the material absorbs light, the attenuation of the light can be described by defining a complex index of refraction:

$$\tilde{n} = n - ik,$$

where n (the index of refraction) is taken as the real part of the complex index and the **extinction coefficient** k as the imaginary part.

Determining the reflectance when absorption is present involves complex multiplication and therefore the use of complex conjugation:

$$R = \left(\frac{\tilde{n}-1}{\tilde{n}+1}\right)^2 = \left(\frac{\tilde{n}-1}{\tilde{n}+1}\right)\left(\frac{\tilde{n}-1}{\tilde{n}+1}\right)^* = \frac{(n-1)+k^2}{(n+1)+k^2}.$$

4.6.2 Complex Contact Angle

Normally, we consider the values for an angle θ to be real. There are, however, cases in applied problems where we can encounter an angle θ with a complex or imaginary value. Complex trigonometric functions are defined in terms of the complex exponential functions

$$\cos z = \frac{e^{iz} + e^{-iz}}{2}, \quad \sin z = \frac{e^{iz} - e^{-iz}}{2i}$$

By defining the functions in this way, we preserve our familiar trigonometric identities and differentiation formulas. There is one new feature to consider, namely, with this definition it is now possible for the sine and cosine functions to take on values greater than 1. For example, consider $\cos i$:

$$\cos i = \frac{e^{i \cdot i} - e^{-i \cdot i}}{2} = \frac{e^{-1} - e^{1}}{2} = \frac{1}{2e} - \frac{e}{2} = 1.543\ldots$$

Such a situation occurs in material science when determining the hydrophilia or hydrophobia of a surface. **Wettability** is a measure of a liquid's ability to maintain contact with a solid surface. For an ideal, flat surface, the wettability of a liquid is described by its **contact angle**, which is defined as

$$\cos \theta = \frac{Y_{sg} - Y_{sl}}{Y_{sg}},$$

where Y denotes the interfacial tension of the boundary between the phases of solid (s), liquid (l), and gas (g). In experiments where a surface is not ideally flat, contact angle measurements can return values for which

$$\cos \theta > 1.$$

While this equation admits no real solutions for θ, it does admit complex-valued solutions. The inclusion of complex contact angles has led to a more complete model of wettability [15].

4.6.3 The Fibonacci Sequence

The Fibonacci sequence 1, 1, 2, 3, 5, 8, 13 ... appears frequently in popular literature. The sequence was introduced in Section 1.2 and defined by the recurrence relation

$a_0 = a_1 = 1$;

$a_n = (a_{n-1} + a_{n-2})$ for $n \geq 2$. The sequence was defined via a recurrence relation because there is no obvious function that can generate the sequence. If it is assumed that the terms of the Fibonacci sequence are the coefficients of a complex power series expansion

$$f(z) = a_0 + a_1 z + a_2 z^2 + a_3 z^3 + \cdots,$$

then the resulting power series expansion

$$f(z) = \sum_{n=0}^{\infty} a_n z^n = 1 + 1z + \sum_{n=2}^{\infty} (a_{n-1} + a_{n-2}) z^n$$

can be shown to define an analytic function. Distributing z^n in the previous expression gives the result

$$f(z) = 1 + 1z + \sum_{n=2}^{\infty} (a_{n-1} z^n + a_{n-2} z^n).$$

This expression can be rewritten as

$$f(z) = 1 + 1z + \sum_{n=2}^{\infty} a_{n-1} z^n + \sum_{n=2}^{\infty} a_{n-2} z^n.$$

The indexing can be changed to make the expression more transparent: setting $n = k + 2$, we find that

$$f(z) = 1 + 1z + \sum_{k=1}^{\infty} a_{k+1}z^{k+2} + \sum_{k=0}^{\infty} a_k z^{k+2}.$$

The middle two terms of the expression can be combined and a power of z^2 can be pulled out of the right-hand summation to give

$$f(z) = 1 + z\left(1 + \sum_{k=1}^{\infty} a_{k+1}z^{k+1}\right) + z^2 \sum_{k=0}^{\infty} a_k z^k.$$

After a few moments of reflection, we should realize that the previous expression can be rewritten as

$$f(z) = 1 + zf(z) + z^2 f(z).$$

It is now possible to solve for $f(z)$:

$$f(z) - zf(z) - z^2 f(z) = f(z)(1 - z - z^2) = 1.$$

Therefore,

$$f(z) = \frac{1}{1 - z - z^2}.$$

We have managed to define a function whose power series coefficient, when it is taken as a sequence, forms the Fibonacci sequence. Note that the function

$$f(z) = \frac{1}{1 - z - z^2}$$

is not analytic when the denominator is equal to zero, i.e., when $1 - z - z^2 = 0$. Using the quadratic formula to solve for the roots of the equation, we find that

$$z = \frac{-1 \pm \sqrt{(-1)^2 - 4(-1)(1)}}{2(1)} = \frac{1 \pm \sqrt{5}}{2}.$$

Expressing the function in terms of its roots, we get

$$f(z) = \frac{1}{(z - \frac{1+\sqrt{5}}{2})(z - \frac{1-\sqrt{5}}{2})}.$$

It is possible to use the method of partial fraction decomposition (see Section 3.4.6) to find the Maclaurin series expansion of the function $f(z)$. Partial fraction decomposition gives

$$f(z) = \frac{1}{(z - \frac{1+\sqrt{5}}{2})(z - \frac{1-\sqrt{5}}{2})} = \frac{-1}{\sqrt{5}(z - \frac{1+\sqrt{5}}{2})} + \frac{1}{z - \frac{1-\sqrt{5}}{2}}.$$

Letting

$$r_1 = \frac{1 - \sqrt{5}}{2} \text{ and } r_2 = \frac{1 + \sqrt{5}}{2},$$

we find that

$$f(z) = \frac{-1}{\sqrt{5}(z - r_1)} + \frac{1}{\sqrt{5}(z - r_2)}.$$

This expression can be rearranged to make it more apparent that each of the terms is the sum of a geometric series:

$$f(z) = \frac{1}{r_1\sqrt{5}(1 - \frac{z}{r_1})} - \frac{1}{r_2\sqrt{5}(1 - \frac{z}{r_2})} = \frac{1}{r_1\sqrt{5}} \sum_{n=0}^{\infty} (\frac{z}{r_1})^n + \frac{1}{r_2\sqrt{5}} \sum_{n=0}^{\infty} (\frac{z}{r_2})^n.$$

The summations can be combined to obtain the expression

$$f(z) = \sum_{n=0}^{\infty} \frac{1}{\sqrt{5}} \left[\frac{r_1^{n+1}}{(-1)^{n+1}} - \frac{r_2^{n+1}}{(-1)^{n+1}} \right] z^n.$$

Substituting our expressions for r_1 and r_2 into this equation gives the result

$$f(z) = \sum_{n=0}^{\infty} \frac{1}{\sqrt{5}} \left[\left(\frac{1 + \sqrt{5}}{2}\right)^{n+1} - \left(\frac{1 - \sqrt{5}}{2}\right)^{n+1} \right] z^n = \sum_{n=0}^{\infty} a_n z^n.$$

Note that the Maclaurin series coefficients

$$a_n = \frac{1}{\sqrt{5}} \left[\left(\frac{1 + \sqrt{5}}{2}\right)^{n+1} - \left(\frac{1 - \sqrt{5}}{2}\right)^{n+1} \right]$$

form the Fibonacci sequence. In short, we have used our knowledge of complex series expansions to find a function that gives the Fibonacci sequence in terms of the formula

$$a_n = \left\{ \frac{1}{\sqrt{5}} \left[\left(\frac{1 + \sqrt{5}}{2}\right)^{n+1} - \left(\frac{1 - \sqrt{5}}{2}\right)^{n+1} \right] \right\},$$

rather than the recurrence relationship given in Section 1.2.

4.6.4 Legendre Polynomials

Legendre polynomials appear frequently in applied problems involving spherical symmetry, for example, in problems involving potentials. This example of a complex power series expansion is interesting because the coefficients of the series expansion are *functions, rather than constants*. The functions that appear as the coefficients of the Maclaurin series expansion of the function

$$f(x) = \frac{1}{\sqrt{1 - 2az + z^2}} = \sum_{n=0}^{\infty} P_n(a) z^n \qquad (4.18)$$

are known as the Legendre polynomials $P_n(a)$.

To find the Maclaurin series expansion of this function, we write

$$f(x) = \frac{1}{\sqrt{1 - 2az + z^2}} = \sum_{n=0}^{\infty} a_n z^n,$$

where

$$\sum_{n=0}^{\infty} a_n z^n = f(0) + \frac{f'(0)}{1!} z + \frac{f''(0)}{2!} z^2 + \frac{f'''(0)}{3!} z^3 + \cdots$$

and the coefficient of the series expansion are given by the formula

$$a_n = \frac{f^{(n)}(0)}{n!}.$$

Expressed in English, the coefficients of the series expansion are obtained by successively differentiating the function of interest with respect to z and then evaluating the derivatives at the origin and dividing by $n!$.

The first term of the Maclaurin series is found by simply evaluating the function at $z = 0$:

$$f(0) = \sqrt{1 - 2a(0) + 0^2} = 1,$$

so that

$$a_0 = 1.$$

The second coefficient of the series is found by differentiating the function once:

$$f'(x) = (1 - 2az + z^2)^{\frac{3}{2}}(a - z),$$

then evaluating the result at the origin, which gives

$$f'(0) = (1 - 2a0 + 0^2)^{\frac{3}{2}}(a - 0) = a.$$

Finally, dividing by 1!, we obtain the coefficient

$$a_1 = \frac{a}{1!}.$$

For the third coefficient we differentiate the function twice and evaluate at $z = 0$, to get

$$f''(0) = 3a^2 - 1.$$

Then, dividing by 2!, we find that

$$a_3 = \frac{3a^2 - 1}{2!}.$$

Repeating the process one more time, we find that the fourth coefficient of the series is

$$a_3 = \frac{15a^3 - 9a}{3!} = \frac{5a^3 - 3a}{2},$$

so that the Maclaurin series expansion looks like

$$f(x) = 1 + \frac{a}{1}z + \frac{3a^2 - 1}{2}z^2 + \frac{5a^2 - 3a}{2}z^3 + \cdots.$$

As we will discover in Section 5.3, the coefficients of the series are the Legendre polynomials

$$P_0(a) = 1, \tag{4.19}$$

$$P_1(a) = a, \tag{4.20}$$

$$P_2(a) = \frac{3a^2 - 1}{2}, \tag{4.21}$$

and

$$P_3(a) = \frac{5a^3 - 3a}{2}, \tag{4.22}$$

which are solutions of Legendre's differential equation.

4.6.5 Complex Series: Diffraction Grating

Oscillations are frequently represented using complex notation, and electro-magnetic oscillations (light waves or radio waves) are one such example. Consider a set of equally spaced radiation sources. The sources could be a set of set of equally spaced antenna, as in a phased array radar, or a set of equally spaced transparent slits, as in a diffraction grating.

A diffraction grating is an optical element used in spectroscopy. The periodic structure of a grating diffracts incoming light and disperses the light into its constituent spectra. When used in transmission, a diffraction grating can be modeled as a series of N equally spaced slits. If the N slits were uniformly illuminated, then each slit would contribute the same amount of light (a) to the total amplitude. However, the contribution from each slit would also be phase shifted by a small amount δ because each slit is offset from the neighboring slits. The complex amplitude of the light transmitted by N slits would be the sum of the light transmitted through each of the equally spaced (phase shifted) slits, so that the total transmitted amplitude A is given by

$$A = a + ae^{i\delta} + ae^{i2\delta} + ae^{i3\delta} + \cdots + ae^{iN\delta}.$$

We would need to sum this series in order to determine the total amplitude of the transmitted light. Note that the finite series is a geometric series with the common ratio $r = e^{i\delta}$:

$$A = a(1 + e^{i\delta} + e^{i2\delta} + e^{i3\delta} + \cdots + e^{iN\delta}).$$

In Chapter 2, we determined that the nth partial sum (S_n) of a geometric series is given by the expression

$$S_n = \frac{a(1 - r^n)}{1 - r},$$

so the total transmitted amplitude of N equally illuminated slits is

$$A = a\frac{1 - e^{iN\delta}}{1 - e^{i\delta}}.$$

4.6.6 Laurent Series Expansion

As a final example, consider the series expansions (both the Laurent and Maclaurin series expansions) of the function

$$f(z) = \frac{e^z}{z^2 + 1}.$$

Note that at the origin, the given function is analytic, so that the point $z = 0$ is a regular point. This means that we can expand the function in a Maclaurin series around the origin.

Now note that the given function has poles of order 2 at both $z = i$ and $z = -i$. In other words, the Maclaurin series expansion will be valid for $z < |1|$, and the Laurent series expansion will be valid for $z > |1|$.

The Maclaurin series expansion is found by simply substituting in the series expansion for both the exponential function and the geometric sum term:

$$\frac{e^z}{z^2 + 1} = \frac{e^z}{1 - (-z^2)} = (1 + z + \frac{z^2}{2!} + \frac{z^3}{3!} + \cdots)(1 - z^2 + z^4 - z^6 + \cdots).$$

Completing the first few multiplications, we obtain the Maclaurin series expression

$$\frac{e^z}{z^2 + 1} = 1 + z - \frac{1}{2}z^2 - \frac{5}{6}z^3 + \frac{13}{24}z^4 + \cdots = \sum_{n=-0}^{\infty}\sum_{m=0}^{\infty}(-1)^m \frac{z^{n+2m}}{n!}, \text{ for } |z| < 1.$$

For the Laurent series expansion in the case where $z > |1|$, we need to note that the series expansion for the exponential function is valid for all z. Therefore, we still have

$$e^z = (1 + z + \frac{z^2}{2!} + \frac{z^3}{3!} + \cdots).$$

However, the geometric series expansion of the function,

$$\frac{1}{1 - (-(z^2))} = 1 - z^2 + z^4 + \cdots = \sum_{n=0}^{\infty}(-(z^2))^n,$$

will not converge for $z > |1|$. It is possible to rewrite this function and expand it as a geometric series in powers of $1/z^2$, i.e.,

$$\frac{1}{z^2(1 + \frac{1}{z^2})} = \frac{1}{z^2}\sum_{n=0}^{\infty}\frac{(-1)^m}{z^{2m}},$$

in which case the Laurent series expansion of the given function for $z > |1|$ is given as

$$\frac{e^z}{z^2+1} = \frac{e^z}{z^2(1+\frac{1}{z^2})} = \frac{1}{z^2}\sum_{n=0}^{\infty}\frac{z^n}{n!}\sum_{m=0}^{\infty}\frac{(-1)^m}{z^{2m}} \quad \text{for } z > |1|.$$

After some tedious multiplication and collecting of like terms, it is possible to express the Laurent series expansion in the form shown in Section 4.5:

$$\frac{e^z}{z^2+1} = \left(\frac{1}{2}-\frac{1}{4}+\cdots\right) + z\left(\frac{1}{3!}-\frac{1}{5!}+\cdots\right) + z^2\left(\frac{1}{4!}+\cdots\right) + z^3\left(\frac{1}{5!}+\cdots\right) + \cdots$$

$$-\frac{1}{z}\left(1-\frac{1}{3!}+\cdots\right) - \frac{1}{z^2}\left(1-\frac{1}{2}+\cdots\right) - \frac{1}{z^3}\left(1-\frac{1}{3!}+\cdots\right) - \cdots$$

5

Series Solutions for Differential Equations

5.1 Introduction

A differential equation is an equation that relates an unknown function to a number of its derivatives, for example,

$$y''(x) = xy(x).$$

Differential equations occur in applied problems that concern rates, for example, the change in an object's position with respect to time (dx/dt), the change in temperature with respect to the distance from a heat source (dT/dx), or the change in concentration with respect to the volume (dC/dV). If an equation only contains ordinary derivatives, it is called an **ordinary differential equation**. The equations

$$\frac{d^2y}{d^2t} = -g,$$

$$f'(x) = 3x^2 - 4x + 5,$$

and

$$\frac{d^3y}{d^3x} + 5y = 0$$

are examples of ordinary differential equations. An equation is a **partial differential equation** when it contains only partial derivatives, for example,

$$\frac{\partial u}{\partial x} = \frac{\partial u}{\partial y},$$

$$\frac{\partial u}{\partial t} = c\frac{\partial^2 u}{\partial^2 x},$$

142

and

$$\frac{\partial^2 u}{\partial x \partial y} = 5.$$

Differential equations are further classified as **linear, nonlinear, homogeneous**, and **nonhomogeneous**. These equations are also classified by their **order** of differentiation: **first order, second order**, ..., **nth order**. The order of a differential equation is that of the highest-order derivative in the equation. To gain a better sense of these classifications, let us classify the three previous examples of ordinary differential equations:

$\frac{d^2 y}{d^2 t} = -g$ (second-order, linear, nonhomogenous),

$f'(x) = 3x^2 - 4x + 5$ (first-order, nonlinear, nonhomogeneous),

and

$$\frac{d^3 y}{d^3 x} + 5y = 0 \quad \text{(third-order, linear, homogenous)},$$

and the following examples of partial differential equations:

$\frac{\partial u}{\partial x} = \frac{\partial u}{\partial y}$ (first-order in x and y),

$\frac{\partial u}{\partial t} = c\frac{\partial^2 u}{\partial^2 x}$ (first-order in t, second-order in x),

and

$$\frac{\partial^2 u}{\partial x \partial y} = 5 \quad \text{(second-order)}.$$

First- and second-order linear differential equations occur frequently in physical and/or applied problems. The techniques for solving the various classes of these equations include separating variables, exact integration, changing the variables, the method of undetermined coefficients, Laplace transforms, and more. These solution methods are detailed in books on differential equations and are beyond the scope of a text on infinite series and sequences, except for one solution technique, series solutions for differential equations, which does fall within the scope of this text.

5.2 Series Solutions for Differential Equations

Many differential equations have solutions that cannot be expressed in terms of elementary or known functions. For example, the simple-looking differential equation

$$y'' - xy$$

has solutions that are not algebraic functions or that cannot be expressed as simple combinations of exponential or trigonometric functions [16]. Linear, homogenous differential equations of the form

$$y'' + f(x)y' + g(x)y = 0 \tag{5.1}$$

occur frequently in physical problems. One method of solving these equations is to assume a power series solution of the form

$$y(x) = \sum_{n=0}^{\infty} A_n (x - x_0)^n.$$

By substituting an assumed infinite series expression into the differential equation, we can obtain a power series expansion that converges to the solutions $y(x)$, *if such solutions exist*. However, before we proceed any further with developing the series solution method, we first need to define some terms.

A **linear, homogeneous differential equation** is an equation of the form

$$a_0 y^{(n)}(x) + a_1 y^{(n-1)}(x) + \cdots + a_{n-1}y'(x) + a_n y(x) = 0, \tag{5.2}$$

where the coefficients a_n are either constants or functions of x. A linear differential equation is an equation whose solutions can be linearly combined to create additional solutions. While a homogenous equation is one in which every term in the equation contains $y(x)$ or a derivative of $y(x)$, an **inhomogeneous** equation

$$a_0 y^{(n)}(x) + a_1 y^{(n-1)}(x) + \cdots + a_{n-1}y'(x) + a_n y(x) = c, \tag{5.3}$$

in contrast, includes a term that does not depend on y.

When is it plausible to expect a power series solution for a linear homogenous differential equation? To answer this question, consider the differential equation

$$y^{(n)} + a_1(x)y^{(n-1)} + \cdots + a_{n-1}(x)y' + a_n(x)y = 0,$$

in which the coefficients $a_n(x)$ are functions of x. Such a linear homogenous differential equation is said to have an **ordinary point** at x_0 if each function $a_n(x)$ is **analytic** at x_0. A point that is not ordinary is a **singular point**. We first encountered singular points in Section 4.5, where we defined singular points or singularities as points where a function is either not differentiable or not single-valued.

Recall that if the functions $a_n(x)$ are analytic at a point x_0, then they can be represented by their Taylor series expansions within some interval of convergence around x_0. This would then suggest that the differential equation

$$y^{(n)} + a_1(x)y^{(n-1)} + \cdots + a_{n-1}(x)y' + a_n(x)y = 0$$

is analytic at the point x_0 and that the analytic solutions $y(x)$ to the differential equation can be represented by their Taylor series expansions around x_0. To summarize, for differential equations of the form

$$y'' + f(x)y' + g(x)y = 0$$

in which both $f(x)$ and $g(x)$ are analytic at x_0, it is possible to assume a series solution in the form of a power series

$$y(x) = \sum_{n=0}^{\infty} A_n(x - x_0)^n.$$

Substituting the assumed power series expression into the differential equation and solving gives the **recursion relations** that determine the series coefficients A_n for $A_n \geq A_2$. In other words, we have obtained a power series expression for the solutions of the differential equation. It may occur to you to ask: Why are the two coefficients A_0 and A_1 undetermined? Well, recall that the general solution to a second-order differential equation will contain two arbitrary constants. In the series solution method, the arbitrary constants are the coefficients A_0 and A_1. These coefficients correspond to the initial values of the function and its first derivative

$$A_0 = y(x_0) \text{ and } A_1 = y'(x_0),$$

and the values of these coefficients would be determined by the initial conditions of the problem.

As an example, consider solving the equation

$$(x + 1)y'' + y' + 2y = 0.$$

Rewriting the equation so that it appears in the desired form,

$$y'' + f(x)y' + g(x)y = 0,$$

we obtain the result

$$y'' + \frac{1}{x+1}y' + \frac{2}{x+1}y = a_0(x)y'' + a_1(x)y' + a_2(x)y = 0.$$

With this form, it is easier to identify the singular points of the equation. On inspection, we can see that the equation is singular or has a singular point at $x = -1$ and that all other points are ordinary. It is around an ordinary point of a differential equation that we can seek a power series solution. As the point $x = 0$ is an ordinary point of this differential equation, we can try to find a power series expression of the form

$$y(x) = \sum_{n=0}^{\infty} A_n x^n.$$

That is, we can seek a Maclaurin series solution, which is an expansion around the origin. Expanding the functions $a_n(x)$ in their Maclaurin series representation, we obtain

$$a_0 = 1 a_0 = 1 \qquad\qquad\qquad \text{for all } x,$$

$$a_1(x) = \tfrac{1}{1+x} = \tfrac{1}{1-(-x)} = \sum_{n=0}^{\infty} (-x)^n \quad \text{for } |x| < 1,$$

$$a_2(x) = \tfrac{2}{1+x} = \tfrac{2}{1-(-x)} = 2\sum_{n=0}^{\infty} (-x)^n \quad \text{for } |x| < 1.$$

This means that the Maclaurin series representation of the solutions for the given differential equation will converge, at least within the interval $|x| < 1$.

To determine the coefficients A_n of the series solutions, we substitute the series expression into the original differential equation and differentiate the terms of the power series expression the required number of times. The first and second derivatives of the power series are given by

$$y'(x) = \sum_{n=1}^{\infty} n A_n x^{n-1}$$

and

$$y''(x) = \sum_{n=2}^{\infty} n(n-1) A_n x^{n-2}.$$

Substituting these expressions into the original, the differential equation becomes

$$\sum_{n=2}^{\infty} n(n-1)A_n x^{n-1} + \sum_{n=2}^{\infty} n(n-1)A_n x^{n-2} + \sum_{n=1}^{\infty} n A_n x^{n-1} + 2\sum_{n=0}^{\infty} A_n x^n = 0.$$

Thus, the original differential equation is now expressed as the sum of four separate power series. These series need to be combined (summed) in a fashion that allows them to be brought together under a single summation sign ($\sum$). We can do this by combining like terms from each series, i.e., combining the terms that have the same exponents for x.

Note that the first series begins with a term that has x raised to the power 1, while the last three series begin with terms that are constants, i.e., with x raised to the power 0. Collecting like terms and grouping them together, we obtain the expression

$$(2A_2 + A_1 + 2A_0) + \sum_{n=2}^{\infty} n(n-1)A_n x^{n-1} + \sum_{n=3}^{\infty} n(n-1)A_n x^{n-2}$$

$$+ \sum_{n=2}^{\infty} nA_n x^{n-1} + 2\sum_{n=1}^{\infty} A_n x^n = 0.$$

Note that each of the four series now starts with a term that has x raised to the power 1. In order to bring these four series under a single summation sign, the consecutive terms of each series need to have the same exponents. To accomplish this, we change or shift the indices of the series so that the exponent of x is the same for each series. In this example, each exponent will be made equal to $k - 2$. Setting $n = k - 1$ in the first series, we obtain the expression

$$\sum_{k=3}^{\infty} (k-1)(k-2)A_{k-1}x^{k-2}.$$

For the second series, setting $n = k$ gives the result

$$\sum_{k=3}^{\infty} k(k-1)A_k x^{k-2}.$$

For the third series, setting $n = k - 1$ gives

$$\sum_{k=3}^{\infty} (k-1)A_{k-1}x^{k-2}.$$

Finally, for the fourth series, setting $n = k - 2$ gives

$$2\sum_{k=3}^{\infty} A_{k-2}x^{k-2}.$$

So the expression for the series solution becomes

$$(2A_2 + A_1 + 2A_0) + \sum_{k=3}^{\infty} (k-1)(k-2)A_{k-1}x^{k-2} + \sum_{k=3}^{\infty} k(k-1)A_k x^{k-2}$$

$$+ \sum_{k=3}^{\infty} (k-1)A_{k-1}x^{k-2} + \sum_{k=3}^{\infty} A_{k-2}x^{k-2} = 0,$$

and it is now possible to bring all four of the series together under a single summation sign:

$$(2A_2 + A_1 + 2A_0) + \sum_{k=3}^{\infty} [(k-1)(k-2)A_{k-1} + k(k-1)A_k + (k-1)A_{k-1}$$

$$+ 2A_{k-2}]x^{k-2} = 0.$$

This expression is the Maclaurin series representation of the solutions for the original differential equation,

$$(x+1)y'' + y' + 2y = 0.$$

While we have found a series expansion of the solutions for our differential equation, we have yet to determine the recursion relations between the coefficients A_n. To do this, we need to recognize that in order for a power series expression,

$$B_0 + B_1 x + B_2 x^2 + \cdots = 0,$$

to be identically zero for all x, each coefficient B_n must be identically zero. Therefore, setting the first coefficient of the series equal to zero, we obtain the expression

$$2A_2 + A_1 + 2A_0 = 0.$$

Similarly, the remaining coefficients are given by

$$(k-1)(k-2)A_{k-1} + k(k-1)A_k + (k-1)A_{k-1} + 2A_{k-2} = 0, \quad \text{for } k \geq 3.$$

This expression can be simplified to

$$k(k-1)A_k + (k-1)(k-3)A_{k-1} + 2A_{k-2} = 0, \quad \text{for } k \geq 3,$$

which gives us the recursion relations for the coefficients A_n of the series. Note that the coefficient A_2 is defined in terms of A_0 and A_1:

$$A_2 = -\frac{1}{2}A_1 - A_0,$$

where A_1 and A_0 are arbitrary constants corresponding to the initial value of the function and its first derivative, respectively, i.e., $A_0 = y(0)$ and $A_1 = y'(0)$. The coefficients A_n are expressed in terms of A_{n-1} and A_{n-2}:

$$A_k = -\frac{(k-3)}{k}A_{k-1} - \frac{2}{k(k-1)}A_{k-2}, \quad \text{for } k \geq 3.$$

For $k = 3$, we find that

$$A_3 = -\frac{1}{3}A_1,$$

that is, A_3 is determined from the value of A_1. For $k = 4$, we find that

$$A_4 = -\frac{1}{4}A_3 - \frac{1}{6}A_2.$$

If we now substitute our previous expressions for A_3 and A_2, we obtain

$$A_4 = -\frac{1}{4}\left(\frac{-1}{3}\right)A_1 - \frac{1}{6}\left(\frac{-1}{2}A_1 - A_0\right) = \frac{1}{6}A_1 + \frac{1}{6}A_0,$$

so that the first few terms of the Maclaurin series representation of the solutions are

$$y(x) = A_0 + A_1 x + \left(-\frac{1}{2}A_1 - A_0\right)x^2 + \left(-\frac{1}{3}A_1\right)x^3 + \left(\frac{1}{6}A_1 + \frac{1}{6}A_0\right)x^4 + \cdots$$

Separating terms with a common factor A_0 from those with a common factor A_1 gives us

$$y(x) = A_0\left(1 - x^2 + \frac{1}{6}x^4 + \cdots\right) + A_1\left(x - \frac{1}{2}x^2 - \frac{1}{3}x^3 + \frac{1}{6}x^4 \cdots\right).$$

If we let each series represent a function,

$$y_1(x) = \left(1 - x^2 + \frac{1}{6}x^4 + \cdots\right)$$

and

$$y_2(x) = \left(x - \frac{1}{2}x^2 - \frac{1}{3}x^3 + \frac{1}{6}x^4 \cdots\right),$$

then the general solution for the original differential equation has the form

$$y(x) = A_0 y_1(x) + A_1 y_2(x),$$

where the values of the arbitrary constants A_0 and A_1 are set by the initial conditions and the series solutions $y_1(x)$ and $y_2(x)$ converge, at least within the interval $|x| < 1$.

We end this section with some final comments about seeking series solutions for differential equations. In general, it will be difficult to "sum" the series solutions to find a closed form expression for the solutions. However, if we are able to sum the series, i.e., recognize the series as the expansion of an elementary function, then the differential equation can be solved by some other means without needing to resort to a series solution. Moreover, it is important to keep in mind that if power series solutions exist, the series solution method will find those solutions, but if the solutions cannot be represented in a power series, this method will not identify the solutions (recall from Chapter 3 that not all functions, i.e., solutions, have power series expansions).

5.3 Generalized Series Solutions and the Method of Frobenius

For the sake of telling the story in Section 5.2, we neglected a fine detail. If we now address this neglected detail, it will be possible to generalize the series solution method for differential equations and broaden its applicability. In the previous section, we developed a series expansion method capable of finding analytic solutions for the differential equation

$$y'' + f(x)y' + g(x)y = 0$$

around *ordinary points* of the differential equation, i.e., at points where the functions $f(x)$ and $g(x)$ are analytic. We did not consider the possible existence of analytic solutions at points where the differential equation is singular. Our current series method will not work for these cases, as the solutions we would be seeking cannot be expanded in a power series of the form

$$y(x) = \sum_{n=0}^{\infty} A_n (x - x_0)^n.$$

To better understand why it is not possible to find a simple power series expansion at a singular point, recall that the coefficients A_n of a power series are found using a Taylor series expansion

$$A_n = \frac{f^{(n)}(x)}{n!}.$$

If, at some point x, the functions $f(x)$ and $g(x)$ in the equation

$$y'' + f(x)y' + g(x)y = 0,$$

are not analytic (infinitely differentiable) – i.e., if they are singular at x – then they do not have a Taylor series expansion, and it is therefore not possible to represent or expand the differential equation in terms of the coefficients A_n of a power series.

However, the occurrence of a singular point in a differential equation does not preclude the existence of analytic solutions at that point. In such cases, it may be possible to represent the solutions of the differential equation in an infinite series expansion at a singular point x_0 using a series expansion of the form

$$y(x) = (x - x_0)^s \sum_{n=0}^{\infty} A_n(x - x_0)^n = \sum_{n=0}^{\infty} A_n(x - x_0)^{n+s}, \text{ for } A_0 \neq 0. \quad (5.4)$$

In such a case, the exponent s is not necessarily an integer or even real; the exponent s can be complex, positive, negative, and/or a fraction. Such a series expansion is referred to as a **generalized power series,** and the method for finding such solutions is known as the **method of Frobenius.**

As an example, consider the differential equation

$$x^2 y'' - 2xy' + 2y = 0.$$

By rewriting the equation as

$$y'' - \frac{2}{x} y' + \frac{2}{x^2} y = 0,$$

we can see that the differential equation is singular at the point $x = 0$; yet the general solution to the differential equation can be shown to be

$$y(x) = A_1 x + A_2 x^2,$$

which is analytic at $x = 0$. That is, every solution is analytic at the point where the differential equation is singular. This is the case for a class of differential equations of the form

$$(x - x_0)^2 y'' + (x - x_0)F(x)y' + G(x)y = 0.$$

Such an equation is said to have a **regular singular point** or a **nonessential singularity** at x_0 if $F(x)$ and $G(x)$ are analytic at x_0. In this case, it is possible to rewrite the differential equation as

$$y'' - \frac{F(x)}{(x - x_0)}y' + \frac{G(x)}{(x - x_0)^2}y = 0.$$

If we let

$$f(x) = \frac{F(x)}{x - x_0}$$

and

$$g(x) = \frac{G(x)}{(x - x_0)^2},$$

then we obtain the equation

$$y'' + f(x)y' + g(x)y = 0.$$

Differential equations of this form are said to be **regular** (have a **nonessential singularity**) at x_0 when the functions

$$(x - x_0)f(x)$$

and

$$(x - x_0)^2 g(x)$$

are both analytic at x_0, that is, the functions can be expanded in power series representations of the forms

$$(x - x_0)f(x) = \sum_{n=0}^{\infty} f_n x^n$$

and

$$(x - x_0)^2 g(x) = \sum_{n=0}^{\infty} g_n x^n.$$

In such cases, we can seek solutions for the second-order differential equation in the form of a generalized powers series

$$y(x) = (x - x_0)^s \sum_{n=0}^{\infty} A_n (x - x_0)^n. \tag{5.5}$$

We shall refer to this generalized power series expression as the "Frobenius" series solution for the differential equation.

As an example, let us use the method of Frobenius to find the series solutions for a second-order differential equation of the form

$$x^2y'' + xF(x)y' + G(x)y = 0.$$

For the sake of transparency, we will assume that both $F(x)$ and $G(x)$ are analytic at the origin and that we know their Maclaurin series expansions on some interval of convergence:

$$F(x) = \sum_{n=0}^{\infty} f_n x^n \text{ and } G(x) = \sum_{n=0}^{\infty} g_n x^n, \quad \text{for } |x| < r.$$

We make these assumptions so that we can focus on the method of Forbenius and not be distracted by superfluous details. With our given assumptions, the differential equation

$$x^2y'' + xF(x)y' + G(x)y = 0$$

becomes regular, or has a nonessential singular point, at the origin. This means that we can seek a generalized series solution of the form

$$y(x) = x^s \sum_{n-0}^{\infty} a_n x^n = \sum_{n=0}^{\infty} a_n x^{n+s}.$$

In this case, the first and second derivatives of the assumed series solution are

$$y'(x) = \sum_{n=0}^{\infty} (n+s)a_n x^{n+s-1}.$$

and

$$y''(x) = \sum_{n=0}^{\infty} (n+s)(n+s-1)a_n x^{n+s-2}.$$

Furthermore,

$$xF(x)y'(x) = x^s \left(\sum_{n=0}^{\infty} (n+s)a_n x^n \right) \left(\sum_{n=0}^{\infty} f_n x^n \right),$$

which can be rewritten as

$$xF(x)y'(x) = \sum_{n=0}^{\infty} \left(\sum_{k=0}^{n} (k+s)a_k x^k f_{n-k} \right) x^{n+s}$$

(see Section 3.4.2), while the expression for the $G(x)y(x)$ term can be written as

$$G(x)y(x) = x^s \sum_{n=0}^{\infty} g_n x^n \sum_{n=0}^{\infty} a_n x^n = \sum_{n=0}^{\infty} \left(\sum_{k=0}^{n} a_k g_{n-k} \right) x^{n+s}.$$

Substituting our various series expressions into the original differential equation and combining like terms, i.e., those with the same powers of x, we obtain the expression

$$\sum_{n=0}^{\infty} \left((n+s)(n+s-1)a_n + \sum_{k=0}^{n} \left((k+s)f_{n-k} + g_{n-k} \right) a_k \right) x^{n+s} = 0.$$

This expression is the series expansion of the solutions for the original differential equation.

We now need to determine the recursion relations for the series coefficients a_n. Note that for the preceding equation to be identically zero for all values of x, the coefficients for each power of x must all be equal to zero. The coefficient for the first term of the series expansion, for $n = 0$, is

$$[s(s-1) + (sf_0 + g_0)]a_0 = 0.$$

If we assume that the coefficient a_0 is not zero – after all, $a_0 x^s$ represents the first term of our series solution – we can conclude that

$$s(s-1) + sf_0 + g_0 = 0,$$

which can be rewritten as

$$s^2 + (f_0 - 1)s + g_0 = 0. \tag{5.6}$$

This equation is called the **indicial equation**, and the allowed values of s correspond to the roots s_1 and s_2 of the indicial equation (note that when determining the roots of a quadratic equation, it is possible for s_i to be complex). The two roots s_1 and s_2 are referred to as the **exponents** of the differential equation (at the regular singular point). If the recursion relations for the coefficients a_n for $n \geq 1$ are given by

$$(n+s)(n+s-1)a_n + \sum_{k=0}^{n} \left((k+s)f_{n-k} + g_{n-k} \right) a_k = 0$$

then, collecting all terms with a common factor of a_n, we obtain the recursion relations

$$((n+s)(n+s-1)+(n+s)f_0+g_0)a_n = -\sum_{k=0}^{n-1} \left((k+s)f_{n-k} + g_{n-k} \right) a_k, \quad \text{for } n \geq 1.$$

Recall that the allowed values of s correspond to the value of the two roots s_1 and s_2 of the indicial equation. Thus, we actually have two separate recursion relations, one for each root s_i:

$$\left((n+s_i)(n+s_i-1+(n+s_i)f_0+g_0)\right)a_n = -\sum_{k=0}^{n-1}\left((k+s_i)f_{n-k}+g_{n-k}\right)a_k.$$

From this we can conclude that there are two Frobenius series solutions $S_1(x)$ and $S_2(x)$ of the forms

$$S_1(x) = x^{s_1}\sum_{n=0}^{\infty}a_n x^n \tag{5.7}$$

and

$$S_2(x) = x^{s_2}\sum_{n=0}^{\infty}a_n x^n. \tag{5.8}$$

In order to avoid some tedious mathematics, it is useful to introduce a theorem regarding the form of generalized power series solutions. As we shall see, there are only two possible forms of generalized power series solutions, and the form of the solution is determined by the relative values of the roots s_1 and s_2. Although we will not prove this, **Fuchs's Theorem** states that the generalized power series solutions to a *regular* second-order differential equation will either have the form of two Frobenius series

$$S_1(x) \text{ and } S_2(x),$$

or the form

$$S_1(x) \text{ and } S_1(x)\ln x + S_2(x). \tag{5.9}$$

If the roots (exponents) s_1 and s_2 are real and distinct and the difference between the exponents $s_1 - s_2$ is not a positive integer, there are two Frobenius series solutions:

$$S_1(x) \text{ and } S_2(x).$$

Both series $S_i(x)$ converge, at least for $|x| < r$, and are solutions of the differential equation, at least within the interval $(0, r)$.

If the exponents s_1 and s_2 are equal or if they are real and distinct and the difference between the exponents $s_1 - s_2$ *is* a positive integer, there are two solutions,

$$S_1(x) \text{ and } S_1(x)\ln x + S_2(x),$$

where both series $S_i(x)$ are Frobenius series and converge, at least for $|x| < r$, and the solutions for the differential equation converge, at least within the interval $(0, r)$.

5.4 Introduction to Special Functions: Bessel, Hermite, and Legendre

The method of Frobenius allows us to solve a broad class of differential equations of the form

$$y'' + f(x)y' + g(x)y = 0 \qquad (5.10)$$

if the functions

$$xf(x)$$

and

$$xg(x)$$

are analytic, i.e., can be represented by a convergent power series. Many important equations of this form appear in a wide variety of applications. One example is **Legendre's equation,**

$$(1 - x^2)y'' - 2xy' + l(l+1)y = 0. \qquad (5.11)$$

This equation occurs in problems involving spherical symmetry and therefore frequently appears in applied problems involving potentials, such as in electromagnetism, mechanics, and quantum mechanics.

There is a class of polynomials called the **Legendre polynomials** that solve Legendre's equation. As an example, we can find the Legendre polynomials by assuming a series solution and using the method of Frobenius to solve for the series expression. Because of their importance in the physical sciences, Legendre polynomials have been studied extensively, so it is possible to find tables and graphs for these functions in many reference books and textbooks. Another important application of Legendre polynomials is in the expansion or representation of a function in a **Legendre series**. Just as it is possible to represent, or expand, a function $f(x)$ in terms of a power series, i.e., a polynomial expansion of the form

$$f(x) = \sum_{n=0}^{\infty} a_n x^n,$$

it is possible to represent a function in terms of a Legendre series expansion, i.e.,

$$f(x) = \sum_{l=0}^{\infty} c_l P_l(x),$$

where $P_l(x)$ denotes the Legendre polynomials.

Another example of an important differential equation is **Bessel's equation,**

$$x^2 y'' + xy' + (x^2 - p^2)y = 0. \tag{5.12}$$

This equation appears frequently in problems involving cylindrical symmetry, for example, optical and/or angular resolution, the stability of vertical supports, waveguides, modes of vibration of membranes (i.e., drum heads), and the dynamics of floating bodies. Bessel's equation can also be solved by the method of Frobenius; the solutions are known as **Bessel functions.** Just as with Legendre polynomials, Bessel functions have been studied extensively, and you can find entire books on them.

One of the most important cases of Bessel functions occurs when p is an integer; in this case, the Bessel functions can be referred to as cylindrical functions or cylindrical harmonics. A further and very important application of Bessel functions is in the solution of differential equations that do not appear in the standard form

$$x^2 y'' + xy' + (x^2 - P^2)y = 0.$$

In some cases, the solutions of nonstandard differential equations can be expressed in terms of Bessel functions.

The last important set of functions we will consider here are the **Hermite functions** or **Hermite polynomials.** These functions appear as solutions to eigenvalue problems in quantum mechanics. **Hermite's differential equation** has the form

$$y'' - 2xy' + 2py = 0 \tag{5.13}$$

and can be solved by assuming a power series solution. It is interesting to note that when the parameter p in Hermite's equation has an integer value, the series solution method returns a series solution that is finite, not infinite. That is, in such cases, one solution for the differential equation is a polynomial of order n.

As an example, we will solve Legendre's equation,

$$(1 - x^2)y'' - 2xy' + l(l+1)y = 0,$$

using the method of Frobenius. Rewriting the equation as

$$y'' - \frac{2x}{1-x^2}y' + \frac{l(l+1)}{1-x^2}y = 0,$$

we note that it has regular singular points at $x = \pm1$ and that all other points are ordinary. Let us seek a series expansion at the point $x = 1$. It should be noted that Legendre's equation is ordinary at all other points x, so it would be possible to find a power series solution at any of these points (see Section 5.2). That is, we could have chosen to a find a power series solution around some point $x = c$ or even a Maclaurin series solution around the origin $x = 0$ without resorting to the method of Frobenius. However, because we want to demonstrate the method of Frobenius, we will seek a series solution at a regular singular point.

To begin, we rewrite Legendre's equation in a form that will make the resulting series solution easier to write out. In Legendre's equation

$$(1 - x^2)y'' - 2xy' + l(l+1)y = 0,$$

l is simply a constant, so let $l(l + 1) = \lambda$. Furthermore, let us make a change of variables with $r = x - 1$, so that Legendre's equation now looks like

$$(2r + r^2)\frac{d^2y}{d^2r} + 2(1 + r)\frac{dy}{dr} - \lambda y = 0.$$

Assuming a series solution of the form

$$y = r^s \sum_{n=0}^{\infty} a_n r^n,$$

we substitute the series expression into the differential equation. Taking the required number of derivatives and collecting like powers of r, we obtain the expression

$$2s^2 a_0 r^{s-1} + \sum_{n=1}^{\infty} [2(n + s)^2 a_n + [(n + s)(n + s - 1) - \lambda]a_{n-1}]r^{n+s-1} = 0.$$

This equation can be satisfied for all values of r only if the sum of the coefficients for each term of the series equals zero. For the first term of the series, $n = 0$, we obtain the indicial equation

$$2s^2 = 0.$$

This equation has two identical roots, $s_1 = s_2 = 0$, and therefore, by Fuchs's Theorem, there is only one series solution, which has the form

$$y = r^s \sum_{n=0}^{\infty} a_n r^n.$$

Substituting the allowed values of s into the series expression, we obtain

$$\sum_{n=1}^{\infty} [2n^2 a_n + [n(n-1) - \lambda]a_{n-1}]r^{n-1} = 0,$$

so that the recursion relations are given by

$$a_n = \frac{\lambda - n(n-1)}{2n^2} a_{n-1}, \quad \text{for } n \geq 1.$$

Writing out the first few series coefficients, we obtain

$$a_1 = \frac{\lambda - 0}{2(1^2)} a_0,$$

$$a_2 = \frac{\lambda - 1(2)}{2(2^2)} a_1 = \frac{(\lambda - 0)(\lambda - 1(2))}{2^2(1^2)2^2} a_0,$$

so that the series solution has the form

$$y(x) = 1 + \sum_{n=1}^{\infty} \frac{\lambda(\lambda - 1(2))(\lambda - 2(3))...(\lambda - (n-1)n)}{2^n (n!)^2} r^n.$$

Recall that this series is in fact a solution to the differential equation

$$(2r + r^2)\frac{d^2 y}{d^2 r} + 2(1 + r)\frac{dy}{dr} - \lambda y = 0.$$

To obtain a series solution for the original Legendre equation, let $\lambda = l(l+1)$ and let $r = x - 1$. Note that in the cases where l is a nonnegative integer, the series expression will terminate when $n = l$, so that the series expression becomes

$$y(x) = 1 + \sum_{n=1}^{l} \frac{l(l+1)(l(l+1) - 1 \cdot 2)(\lambda - 2 \cdot 3)\cdots(l(l+1) - (n-1)n)}{2^n (n!)^2}(x-1)^n.$$

Functions of this form are known as **Legendre polynomials**, denoted as $P_l(x)$:

$$P_l(x) = 1 + \sum_{n=1}^{l} \frac{l(l+1)(l(l+1) - 1 \cdot 2)(\lambda - 2 \cdot 3)...(l(l+1) - (n-1)n)}{2^n (n!)^2}(x-1)^n.$$

The function $P_l(x)$ is a polynomial of degree l and a solution for the **Legendre differential equation of order l:**

$$(1 - x^2)y'' - 2xy' + l(l+1)y = 0.$$

The first few Legendre polynomials are:

$$P_0 = 1, \tag{5.14}$$

$$P_1 = x, \tag{5.15}$$

$$P_2 = \frac{3}{2}x^2 - \frac{1}{2}, \tag{5.16}$$

$$P_3 = \frac{5}{2}x^3 - \frac{3}{2}x, \tag{5.17}$$

$$P_4 = \frac{35}{8}x^4 - \frac{15}{4}x^2 + \frac{3}{8}, \tag{5.18}$$

$$P_5 = \frac{63}{8}x^5 - \frac{35}{4}x^3 + \frac{15}{8}x. \tag{5.19}$$

Legendre polynomials have the important and useful properties that they are orthogonal, i.e.,

$$\int_{-1}^{1} P_l(x)P_m(x) = 0, \quad \text{for } l \neq m, \tag{5.20}$$

and form a complete set. This means that it is possible to expand functions in an infinite series of Legendre polynomials of the form

$$f(x) = \sum_{n=1}^{\infty} c_l P_l(x); \tag{5.21}$$

such series are referred to as **Legendre series.** We will come back to this topic in Chapter 6.

As a final example, we will use the method of Frobenius to find a series solution for **Bessel's equation of order p:**

$$x^2 y'' + xy' + (x^2 - p^2)y = 0,$$

where p is a nonnegative real constant. This equation has a regular singular point at $x = 0$, so we will assume a series solution of the form

$$y = x^s \sum_{n=0}^{\infty} A_n x^n$$

at the origin. Once again, we substitute the assumed series expansion into the differential equation, perform the required number of differentiations, and collect like powers of x. We obtain the following result:

$$(s(s-1) + s - p^2)A_0 x^s + ((s+1)s + s + 1 - p^2)A_1 x^{s-1}$$
$$+ \sum_{n=2}^{\infty} ([(n+s)(n+s-1) + (n+s) - p^2]A_n + A_{n-2}) x^{n+s} = 0.$$

In order for this to be a solution for all values of x, the sum of all the coefficients for each term of the series must be zero. From the first term of the series, $n = 0$, we obtain the indicial equation

$$s(s-1) + s - p^2 = 0,$$

while the second term of the series, $n = 1$, gives the relation

$$\left((s+1)s + s + 1 - p^2 \right) a_1 = \left((s+1)^2 - p^2 \right) a_1 = 0$$

and the third term of the series gives the recursion relations

$$\left((n+s)^2 - p^2 \right) a_n = -a_{n-2}, \quad \text{for } n \geq 2.$$

Solving for the roots of the indicial equation will give us the allowed values for the exponents s_1 and s_2. Simplifying the indicial equation, we find that

$$s^2 - p^2 = 0.$$

Therefore, the exponents are $s_1 = p$ and $s_2 = -p$. At this point, we now have two separate series to solve, one for each root.

To find the series solution for the case where $s = p$, we substitute the allowed value of the exponent $(s = p)$. For the second term of the series, $n = 1$, we find that

$$\left((s+1)^2 - p^2\right)a_1 = \left((p+1)^2 - p^2\right)a_1 = (2p+1)a_1 = 0,$$

and therefore $a_1 = 0$. Substituting the allowed value of the exponent $(s = p)$ into the third term of the series, we find that

$$n(n + 2p)a_n = -a_{n-2}, \quad \text{for } n \geq 2.$$

In other words, $a_1 = a_3 = a_5 = \cdots = 0$, and the remaining coefficients can be written as

$$a_2 = \frac{-1}{2(2 + 2p)}a_0,$$

$$a_4 = \frac{1}{4(4 + 2p)}a_2 = \frac{1}{2 \cdot 4(2 + 2p)(4 + 2p)}a_0,$$

.
.
.

$$a_{2n} = \frac{(-1)^n}{2^{2n}n!(1+p)(2+p)\ldots(n+p)}a_0.$$

Therefore, the Frobenius series solution has the form

$$y(x) = a_0 x^p \left(1 + \sum_{n=1}^{\infty} \frac{(-1)^n (x/2)^{2n}}{2^{2n}n!(1+p)(2+p)\ldots(n+p)}\right) = J_p(x). \qquad (5.22)$$

Functions of this form are called **Bessel functions of the first kind of order p** and are denoted as $J_p(x)$. The preceding series expression can be simplified and made to look like the standard representation of Bessel functions of the first kind by introducing the gamma function $\Gamma(p + 1)$. However, this is not central to what we are trying to accomplish in this book, and so it will be left for a book on orthogonal polynomials.

We must now consider the solution associated with the second root, $s = -p$. This can be accomplished simply by replacing p with $-p$; there is no need to work through the details of the problem again. In this case, the **Bessel functions of the second kind** are written as

$$J_{-p}(x) = a_0 x^{-p}\left(1 + \sum_{n=1}^{\infty} \frac{(-1)^n (x/2)^{2n}}{2^{2n} n!(1-p)(2-p)\cdots(n-p)}\right). \quad (5.23)$$

Now let us consider Bessel function solutions with Fuchs's Theorem in mind. According to Fuchs's Theorem, when p is not an integer, the Bessel functions of the first and second kind will be independent solutions, and the general solution will be a linear combination of the two Bessel functions:

$$y(x) = AJ_p(x) + BJ_{-p}(x). \quad (5.24)$$

On the other hand, when p is an integer, the Bessel functions of the second kind will not be independent solutions. For integer values of p, the Bessel functions of the second kind will be related to the first solution by the expression

$$J_{-p}(x) = (-1)^p J_p(x).$$

In such cases, the second solution will not be in the form of a Frobenius series, as it will include a term with a logarithm (here the reader may wish to revisit the end of Section 5.2.1).

It is common practice for the second solution to be written as a linear combination of $J_{-p}(x)$ and $J_p(x)$, in which case the general solution appears in the form of a linear combination of the two solutions, for example,

$$y(x) = AJ_p(x) + B[CJ_{-p}(x) + DJ_p(x)].$$

5.5 Examples

In this section, we will present several examples that illustrate how series expansions are used to solve differential equations.

5.5.1 Power Series Solution for $y'' + y = 0$

In this example, we will use the method of power series to solve a differential equation whose solution we know. The utility of this example is that it demonstrates the steps involved in finding a power series solution for a differential equation that is already familiar, which allows us to focus on the method of series solutions.

We know from calculus that the second derivatives of the sine and cosine functions are equal to the negative of the function, i.e.,

$$(\sin x)'' = -\sin x \text{ and } (\cos x)'' = -\cos x,$$

so that the differential equation

$$y'' = -y$$

is true for both the sine and cosine functions. Therefore, the general solution for the differential equation

$$y'' + y = 0$$

is a linear combination of the sine and cosine:

$$y(x) = a \cos x + b \sin x.$$

We will now use the method of power series to solve the same differential equation, i.e.,

$$y'' + y = 0,$$

and show that the series solution converges to the sin and cosine functions. The first step is to assume a series solution of the form

$$y = \sum_{n=0}^{\infty} a_n x^n = a_0 + a_1 x + a_2 x^2 + \cdots$$

Differentiating the power series term by term, we get

$$y' = \sum_{n=1}^{\infty} n a_n x^{n-1} = a_1 + 2a_2 x + 3a_3 x^2 + \cdots$$

Differentiating a second time, we obtain

$$y'' = \sum_{n=2}^{\infty} (n-1) n a_n x^{n-2} = 2a_2 + 2 \cdot 3a_3 x + 3 \cdot 4a_4 x^2 + \cdots$$

Substituting our series expression into the original differential equation gives

$$\sum_{n=2}^{\infty} (n-1) n a_n x^{n-2} + \sum_{n=0}^{\infty} a_n x^n = 0.$$

To combine the two series and bring them under a single summation sign, we need to rewrite the expression so that it is possible to sum terms with like exponents, and to accomplish this we shift or change the indices. If we let $n = n + 2$ in the first series expression, we get

$$\sum_{n=0}^{\infty}(n + 1)(n + 2)a_{n+2}x^n + \sum_{n=0}^{\infty}a_nx^n = 0.$$

Now the two series can be combined to obtain

$$\sum_{n=0}^{\infty}[(n + 1)(n + 2)a_{n+2} + a_n]x^n = 0.$$

For this expression to be true for all values of x, each coefficient must be equal to zero:

$$(n + 1)(n + 2)a_{n+2} + a_n = 0,$$

which leads to the recursion relations given by

$$a_{n+2} = \frac{-a_n}{(n + 1)(n + 2)}.$$

For $n = 0$, we find that

$$a_2 = \frac{-a_0}{1 \cdot 2} = \frac{-a_0}{2!}.$$

For $n = 1$, we obtain

$$a_3 = \frac{-a_1}{2 \cdot 3} = \frac{-a_1}{3!}.$$

For $n = 2$, we get

$$a_4 = \frac{-a_2}{3 \cdot 4} = \frac{-a_0}{1 \cdot 2 \cdot 3 \cdot 4} = \frac{a_0}{4!},$$

and for $n = 3$, we get

$$a_5 = \frac{-a_3}{4 \cdot 5} = \frac{a_1}{1 \cdot 2 \cdot 3 \cdot 4 \cdot 5} = \frac{a_1}{5!}.$$

Note that the coefficients are expressed in terms of either a_0 or a_1, which correspond to the two arbitrary constants in the solution of a second-order differential equation, with a_0 corresponding to the initial value of x and a_1 corresponding to the initial slope of $y(x)$.

Substituting the values of the coefficients back into the power series expression of the solution, we find that

$$y(x) = a_0 + a_1 x - \frac{a_0}{2!}x^2 - \frac{a_1}{4!}x^3 + \frac{a_0}{4!}x^4 + \frac{a_1}{5!}x^5 + \cdots.$$

Collecting like terms, i.e., terms containing a common factor (a_0 or a_1), we get the expression

$$y(x) = \left(a_0 - \frac{a_0}{2!}x^2 + \frac{a_0}{4!}x^4 + \cdots\right) + \left(a_1 x - \frac{a_1}{4!}x^3 + \frac{a_1}{5!}x^5 + \cdots\right).$$

This expression can be further simplified and written as

$$y(x) = a_0\left(1 - \frac{1}{2!}x^2 + \frac{1}{4!}x^4 + \cdots\right) + a_1\left(x - \frac{1}{4!}x^3 + \frac{1}{5!}x^5 + \cdots\right).$$

The first series is recognizable as the power series expansion of cos x and the second series as that of sin x. So the general solution to the differential equation can be written as

$$y(x) = a_0 \cos x + a_1 \sin x.$$

In this example, we used the method of power series to solve a simple differential equation, and this resulted in a series solution that was recognizable and could be written in closed form, i.e., in terms of elementary functions. In general, we would not expect the method of power series to return a series solution that can be expressed in terms of elementary functions. The utility of the power series method is that it can be used to solve problems for which the solution cannot be expressed in terms of known or elementary functions. If, however, the method does produce a solution that can be written in closed form, this suggests that the differential equation could have been solved without resorting to a power series expansion.

5.5.2 Series Solution for the Schrödinger Equation

The one-dimensional Schrödinger equation has the form

$$\frac{\hbar^2}{2m}\frac{\partial^2 \psi}{\partial^2 x} + (E - V(x))\psi = 0, \tag{5.25}$$

where $V(x)$ is the potential describing some interaction. In this example, we will use the method of Frobenius to find a series solution for the Schrödinger equation, assuming a potential of the form

$$V(x) = c\frac{e^{-\alpha x}}{x}.$$

A potential of this form is used to model the interaction between nucleons [17]. Substituting the form of the potential in the Schrödinger equation, we obtain

$$\frac{\hbar^2}{2m}\frac{\partial^2 \psi}{\partial^2 x} + \left(E - c\frac{e^{-\alpha x}}{x}\right)\psi = 0.$$

It is useful to rewrite this equation as

$$\frac{\partial^2 \psi}{\partial^2 x} + \left(A - B\frac{e^{-\alpha x}}{x}\right)\psi = 0,$$

where

$$A = \frac{2Em}{\hbar^2} \quad \text{and} \quad B = \frac{2C}{\hbar^2}.$$

Notice that the equation is singular at the point $x = 0$. So in order to seek a series solution around the origin, we will assume a Frobenius series expansion of the form

$$\psi(x) = \sum_{n=0}^{\infty} a_n x^{n+s}.$$

Substituting the assumed form of the series into the equation and differentiating, we find that

$$\sum_{n=0}^{\infty} a_n(n+s)(n+s-1)x^{n+s-2} + \left(A - B\frac{e^{-\alpha x}}{x}\right)\sum_{n=0}^{\infty} a_n x^{n+s} = 0.$$

Expressing the exponential function in terms of its power series expansion gives the result

$$\sum_{n=0}^{\infty} a_n(n+s)(n+s-1)x^{n+s-2} + \left(A - \frac{B}{x}\sum_{m=0}^{\infty}\frac{-(\alpha m)^m}{m!}\right)\sum_{n=0}^{\infty} a_n x^{n+s} = 0,$$

which can be rewritten as

$$\sum_{n=0}^{\infty} a_n(n+s)(n+s-1)x^{n+s-2} + A\sum_{n=0}^{\infty} a_n x^{n+s} - B\sum_{m=0}^{\infty}\sum_{n=0}^{\infty} a_n x^{n+s+m-1}\frac{-(\alpha)^m}{m!} = 0.$$

The coefficient of the first term of the series ($n = 0$) gives the indicial equation

$$a_0 s(s - 1) = 0.$$

The roots of the equation correspond to the allowed values of the exponent s, which in this case are $s_1 = 1$ and $s_2 = 0$.

Let us consider the first Frobenius series solution, which corresponds to the exponent $s = 1$. In this case, the series solution looks like

$$\sum_{n=0}^{\infty} a_n(n + 1)nx^{n-1} + A\sum_{n=0}^{\infty} a_n x^{n+1} - B\sum_{m=0}^{\infty}\sum_{n=0}^{\infty} a_n x^{n+m} \frac{-(\alpha)^m}{m!} = 0.$$

The coefficient of the second term of the series (for $n = 1$) is

$$a_1 2 - Ba_0 = 0,$$

and therefore

$$a_1 = \frac{Ba_0}{2}.$$

The third coefficient of the series ($n = 2$) is given by the expression

$$a_2(2 + 1)2 + Aa_0 - B(-\alpha a_0 + a_1) = 0,$$

which can be written as

$$a_2 = \frac{-(B\alpha + A)a_0 + Ba_1}{3 \cdot 2} = \frac{-2(B\alpha + A) + B^2}{12} a_0$$

At this point, it is possible to write out the first three terms of the first Frobenius series solution as

$$\psi(x) = a_0\left(x - \frac{Bx^2}{2} + \frac{B^2 - 2(B\alpha + A)}{12}x^3 + \cdots\right).$$

5.5.3 Polynomial Solutions

For some differential equations, the method of power series yields a solution that is not an infinite series but rather a finite one. That is, the series solution terminates at some point because the coefficients of all higher-order terms of the series vanish for some $m > n$. In such cases, the solution for the differential equation is simply a polynomial of the form

$$y(x) = a_0 + a_1 x + a_2 x^2 + \cdots + a_n x^n.$$

In many applied problems, this is a useful and important feature for approximating a solution. For example, in quantum mechanics, the power series solution of Hermite's equation,

$$y'' - 2xy' + 2py = 0 \, ,$$

terminates when $p = n$ in cases where p has integer values. This results in polynomial solutions called **Hermite polynomials,** denoted as

$$H_m(x) = \sum_{n=0}^{m/2} (-1)^n \frac{m!}{n!(m-2n)!} (2x)^{m-2n}. \tag{5.26}$$

As an example of a series solution that terminates, let us solve Hermite's equation. We see that the differential equation has an essential singularity at the point $x = \infty$ and that all other points are ordinary. As the point $x = 0$ is an ordinary point, we will use the power series method to seek the solutions around the origin (i.e., a Maclaurin series).

First, we assume a power series solution of the form

$$y(x) = \sum_{n=0}^{\infty} a_n x^n.$$

We take the first and second derivatives of the power series and substitute the resulting series expressions into Hermite's equation to obtain the formula

$$\sum_{n=2}^{\infty} (n-1)n a_n x^{n-2} - 2x \sum_{n=1}^{\infty} n a_n x^{n-1} + 2p \sum_{n=0}^{\infty} a_n x^n = 0.$$

To combine these separate series, we shift the indices so that the exponent of x in each series is n. For the first series we set $n = n+2$, and for the second series we set $n = n+1$, which gives us

$$\sum_{n=0}^{\infty} (n+1)(n+2)a_{n+2}x^n - 2x \sum_{n=0}^{\infty} (n+1)a_{n+1}x^n + 2p \sum_{n=0}^{\infty} a_n x^n = 0 \, .$$

The expression for the second series can be simplified by bringing the factor of x under the summation sign and shifting the index to obtain

$$\sum_{n=0}^{\infty} (n+1)(n+2)a_{n+2}x^n - 2\sum_{n=0}^{\infty} (n+1)a_n x^n + 2p \sum_{n=0}^{\infty} a_n x^n = 0.$$

These series can now be written under a single summation sign:

$$\sum_{n=0}^{\infty}[(n+1)(n+2)a_{n+2} + 2(p-n)a_n]x^n = 0.$$

For this equation to be satisfied, the coefficient of each power of x must be equal to 0, which results in the recursion relations given by

$$a_{n+2} = \frac{2(p-n)}{(n+1)(n+2)}a_n.$$

In cases where p takes on the value of a nonnegative integer m, at the point where m equals n, the recursion relations give

$$a_{m+2} = a_{m+4} = \cdots = 0.$$

The convention is to set

$$a_0 = \frac{(-1)^{m/2}m!}{(m/2)!}$$

for even m and

$$a_1 = \frac{(-1)^{(m-1)/2}2m!}{[1/2(m-1)]!}$$

for odd m. With these conventions, the expression for the general solution is given by

$$H_m(x) = \sum_{n=0}^{m/2}(-1)^n \frac{m!}{n!(m-2n)!}(2x)^{m-2n}.$$

5.5.4 Series Solutions at Infinity

Up to this point, we have only considered seeking series solutions around a finite point, be it an ordinary point or a regular singular point. In some cases, we may want to know the behavior of a solution in the limit as the variable x becomes infinite. To accomplish this, we simply make the following change of variable:

$$x = \frac{1}{t},$$

so that when t approaches zero from above, x approaches positive infinity, and when t approaches zero from below, x approaches negative infinity.

As an example, consider the differential equation

$$2x^3 \frac{d^2y}{d^2x} + 3x^2 \frac{dy}{dx} - y = 0.$$

If we are interested in finding series solutions that are valid for large values of $|x|$, we make the suggested change of variable and obtain the differential equation

$$2t \frac{d^2Y}{d^2t} + \frac{dY}{dt} - Y = 0.$$

Note that this equation has a regular singular point at $t = 0$, so we can use the method of Frobenius to seek a generalized power series solution at that point. Assuming series solutions of the form

$$Y(t) = \sum_{n=0}^{\infty} a_n t^{n+s}$$

and taking the first and second derivatives of the series expression, we obtain

$$Y'(t) = \sum_{n=0}^{\infty} (n+s) a_n t^{n+s-1}$$

and

$$Y''(t) = \sum_{n=0}^{\infty} (n+s-1)(n+s)\, a_n t^{n+s-2} .$$

Substituting these series expressions into the differential equation, we obtain

$$2\sum_{n=0}^{\infty} (n+s-1)(n+s)a_n t^{n+s-1} + \sum_{n=0}^{\infty} (n+s)a_n t^{n+s-1} - \sum_{n=0}^{\infty} a_n t^{n+s} = 0.$$

The first two series can be combined to give the expression

$$\sum_{n=0}^{\infty} (2n+2s-1)(n+s)a_n t^{n+s-1} - \sum_{n=0}^{\infty} a_n t^{n+s} = 0.$$

Notice that the first series begins with a term involving t^{s-1}, while the second series begins with a term involving t^s. Separating out terms containing a factor of t^{s-1}, we obtain the expression

$$s(2s - 1)a_0 t^{s-1} + \sum_{n=1}^{\infty}(2n + 2s - 1)(n + s)a_n t^{n+s-1} - \sum_{n=0}^{\infty} a_n t^{n+s} = 0.$$

If we now shift the index of the second series by setting $n = n + 1$, we obtain

$$s(2s - 1)a_0 t^{s-1} + \sum_{n=0}^{\infty}(2n + 2s - 1)(n + 1 + s)a_{n+1} t^{n+1} - \sum_{n=0}^{\infty} a_n t^{n+s} = 0.$$

The series summations can now be combined to give the result

$$s(2s - 1)a_0 t^{s-1} + \sum_{n=0}^{\infty}\left[\{(2n + 2s - 1)(n + s)\}a_{n+1} - a_n\right]t^{n+s} = 0.$$

The exponents s_i at $t = 0$ are found by solving for the roots of the indicial equation

$$s(2s - 1) = 0.$$

In this case, the exponents are

$$s_1 = \frac{1}{2} \text{ and } s_2 = 0,$$

so that the independent Frobenius series solutions corresponding to the two exponents are given by

$$Y_1(t) = t^{1/2}\sum_{n=0}^{\infty}\frac{2^n}{(2n + 1)!}t^n$$

and

$$Y_2(t) = \sum_{n=0}^{\infty}\frac{2^n}{(2n)!}t^n.$$

These are, however, the solutions for the transformed equation, not the original equation. To find the solutions for the original equation, we replace t with $1/x$ to obtain

$$y_1(x) = \frac{1}{x^{1/2}}\sum_{n=0}^{\infty}\frac{2^n}{(2n + 1)!}\frac{1}{x^n}$$

and

$$y_2(x) = \sum_{n=0}^{\infty} \frac{2^n}{(2n)!} \frac{1}{x^n},$$

which are the valid (convergent) solutions for large values of $|x|$.

6

Fourier, Legendre, and Fourier-Bessel Series

6.1 Introduction

In Section 3.1, we discussed the fact that an infinite series can be constructed from an infinite sum of functions

$$S(x) = f_1(x) + f_2(x) + f_3(x) + \cdots = \sum_{n=0}^{\infty} f_n(x),$$

and if the infinite series approaches a limit, the limit of the series is a function. Up to this point, we have only considered the possibility that elements of the series are powers of x:

$$f_n(x) = a_n x^n.$$

That is, we have only considered expanding or representing a function in terms of a polynomial of the form

$$S(x) = a_0 + a_1 x + a_2 x^2 + \cdots$$

We have not considered the possibility of representing or approximating a function in terms of an infinite series in which the elements are not necessarily powers of x but other functions, for example,

$$S(x) = a_0 + \sin a_1 x + \sin a_2 x + \sin a_3 x \cdots$$

or

$$\sin(x \sin \theta) = 2 \sum_{n=0}^{\infty} J_{2n+1}(x) \sin(2n+1)\theta,$$

where the series coefficients $J_{2n+1}(x)$ are Bessel functions. One example of the usefulness of such types of series is that they can be used to represent solutions

174

of partial differential equations, just as power series can be used to represent solutions of ordinary differential equations. These types of series expansions are topics worthy of books in themselves, so our coverage of them will be incomplete. We are, however, in the position to introduce and understand their utility and the basic properties of a few such series.

6.2 Fourier Series

While power series can be used to approximate functions, not all functions have a power series representation. Furthermore, there exists an entire class of physical problems that are better solved using other types of series expansions, such as Fourier series. **Real Fourier series** are series whose terms are sine and cosine functions:

$$f(x) = \sum_{k=1}^{\infty} b_k \sin kx + \sum_{k=0}^{\infty} a_k \cos kx, \tag{6.1}$$

where a_k and b_k are real, while the **complex form of the Fourier series** is

$$f(x) = \sum_{n=-\infty}^{n=\infty} c_n e^{inx}. \tag{6.2}$$

To obtain the complex form of the Fourier series, we simply express the sine and cosine functions in terms of complex exponentials:

$$\sin nx = \frac{e^{inx} - e^{-inx}}{2i} \quad \text{and} \quad \cos nx \frac{e^{inx} + e^{-inx}}{2}, \tag{6.3}$$

and substitute these formulas into the expression for the real Fourier series. Because the sine and cosine functions are both periodic functions, Fourier series can only represent periodic functions. Of course, this is a desirable property for physical problems involving oscillations or vibrations.

A Fourier series expansion essentially decomposes a periodic function or signal and then represents the signal in terms of the sinusoidal components (harmonics) present in the signal. When expressed in terms of frequency, the sinusoidal functions of a Fourier series form a harmonic series, so the method is sometimes referred to as **harmonic analysis.** Fourier or harmonic analysis has a broad range of applications across many diverse fields of study, from oceanography and physics to biology and signal processing.

To expand a given periodic function in terms of a series of sines and cosines, we need to identify the period of the function, which is essentially the "interval" over which we are going to find a Fourier series expansion that converges to the function. As both sine and cosine have a period of 2π, it would be natural to use these on an interval of length 2π, e.g., on the interval $(-\pi, \pi)$ or the interval $(0, 2\pi)$.

Unfortunately, not all physical phenomena or problems you will encounter have an interval length (period) of 2π. To accommodate this, the Fourier series expressions can be generalized to an arbitrary interval of length $2l$, e.g., $(-l, l)$ or $(0, 2l)$. While we won't prove it, the Fourier series expansions for a function with a period of $2l$ look like

$$f(x) = \frac{a_0}{2} + \sum_{n=1}^{\infty}(a_n \cos \frac{n\pi x}{l} + b_k \sin \frac{n\pi x}{l}) \qquad (6.4)$$

and/or

$$f(x) = \sum_{n=-\infty}^{n=\infty} c_n e^{\frac{in\pi x}{l}},$$

where the coefficients of the Fourier series are determined by the formulas

$$a_n = \frac{1}{l} \int_{-l}^{l} f(x) \cos \frac{n\pi x}{l} dx, \qquad (6.4)$$

$$b_n = \frac{1}{l} \int_{-l}^{l} f(x) \sin \frac{n\pi x}{l} dx, \qquad (6.5)$$

and

$$c_n = \frac{1}{2l} \int_{-l}^{l} f(x) e^{-\frac{in\pi x}{l}}. \qquad (6.6)$$

It is important to note some significant differences between Fourier series and power series expansions. First, unlike power series, Fourier series can represent functions that are not continuous or differentiable, i.e., functions with discontinuities or "sharp corners." To better understand this distinction, recall that the coefficients of a power series expansion are found by successively differentiating the function being expanded. Therefore, a power series expansion exists only for functions that are infinitely differentiable, whereas many periodic functions are not differentiable – for example, square- or triangular-wave signals.

A second and related difference between power series and Fourier series is that term-by-term differentiation of a convergent power series produces a convergent power series, but this is not the case for some Fourier series. To better understand this difference, consider discontinuous functions. If a function is not differentiable, then neither is its Fourier series representation. If a function is continuous up to the first derivative, i.e., can be differentiated once, then its Fourier series expansion can be differentiated once. If a function is twice differentiable, then its Fourier series expansion its twice differentiable, and so on.

Finally, Fourier series generally do not converge as quickly as power series, so more care must be taken when using Fourier series to approximate solutions.

6.3 Legendre Series

In Section 5.3, we introduced the first few Legendre polynomials:

$$P_0 = 1,$$

$$P_1 = x,$$

and

$$P_2 = \frac{3}{2}x^2 - \frac{1}{2}.$$

The Legendre polynomials $P_l(x)$ are an example of a class of functions known as orthogonal polynomials. While the study of orthogonal polynomials is beyond the scope of this book, one application is not: the representation or expansion of a function in an infinite series of orthogonal polynomials. It is possible to expand functions in a Legendre series in a fashion similar to using a Fourier series, i.e.,

$$f(x) = \sum_{l=0}^{\infty} a_l P_l(x),$$

where the series coefficients a_n are determined by evaluating the integral

$$\int_{-l}^{l} f(x) P_l(l) = a_n \frac{2}{2l+1}$$

– which is similar to the integral used to obtain the coefficients of a Fourier series expansion. Note that the limits of integration for the Legendre coefficients are $x = -1$ and $x = 1$ because Legendre series expansions are valid for the interval $(-1, 1)$. This contrasts with Fourier series expansions, which are valid for an interval of length $2l$, e.g., $(-l, l)$ or $(0, 2l)$.

For any function $f(x)$ that is piecewise smooth on the interval $(-1, 1)$, the Legendre series will converge to $f(x)$ where $f(x)$ is continuous but not necessarily at the endpoints. If $f(x)$ has discontinuities within the interval, then the Legendre series will converge to the midpoint value at those discontinuities. Examples at the end of this chapter will demonstrate this behavior.

6.4 Fourier-Bessel Series

A Fourier-Bessel series expansion is a generalized form of a Fourier series expansion based on Bessel functions rather than on the sine and cosine functions. This generalization is a natural extension of Fourier series expansions when one thinks of Bessel functions as damped sinusoidal functions. Fourier-Bessel series expansions are important for solving partial differential equations in cylindrical coordinates – for example, in the vibrational modes of a drumhead.

The expansion of a function in terms of Bessel functions of the first kind on the interval $(0, 1)$ is written as

$$f(x) = \sum_{n=1}^{\infty} a_n J_n(a_n x), \tag{6.7}$$

where the coefficients of the series are obtained from the integral

$$a_n = \frac{2}{[J_{n+1}(a_n)]^2} \int_0^1 x f(x) J_n(a_n x)\, dx \tag{6.8}$$

and a_n is a zero of the Bessel function, determined by the initial boundary conditions of the problem.

6.5 Examples

In this section, we will provide a few examples of how the various series expansions are used in applied problems.

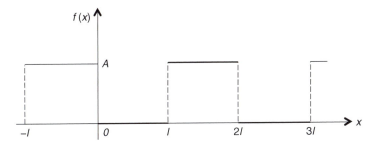

Figure 6.1 A square wave.

6.5.1 Fourier Series Expansion of a Square Wave

As an example of how a Fourier series expansion might be applied in a physical problem, we will expand a square-wave signal (which can be electrical, optical, acoustic, or any other type) of the form shown in Figure 6.1.

Such a square wave can be represented by the function

$$f(x) = \begin{cases} 0, & 0 < x < 1 \\ A, & 1 < x < 2l. \end{cases}$$

It is possible to expand the function $f(x)$ in terms of the complex Fourier series

$$f(x) = \sum_{n=-\infty}^{n=\infty} c_n e^{\frac{in\pi x}{l}}$$

over the interval $(0, 2l)$, where the coefficients c_n are determined by evaluating the integrals

$$c_n = \frac{1}{2l} \int_0^{2l} f(x) e^{-\frac{in\pi x}{l}} = \frac{1}{2l} \int_0^l 0 \cdot e^{-\frac{in\pi x}{l}} + \frac{1}{2l} \int_l^{2l} A \cdot e^{-\frac{in\pi x}{l}}.$$

These integrals are easily evaluated:

$$c_n = 0 + \frac{A}{2l} \frac{e^{-\frac{in\pi}{l}}}{-in\pi l} \bigg|_l^{2l} = \frac{A}{-2in\pi l} (e^{-2in\pi} - e^{-in\pi}) = \frac{A}{-2in\pi} (1 - e^{-in\pi}),$$

where

$$\frac{A}{-2in\pi}(1 - e^{-in\pi}) = \begin{cases} 0, & \text{if } n \neq 0 \text{ is even} \\ \dfrac{A}{-2in\pi}, & \text{if } n \text{ is odd} \end{cases}$$

and

$$c_0 = \frac{A}{2l} \int_l^{2l} dx = \frac{A}{2}.$$

Therefore, the series solution looks like

$$f(x) = \frac{A}{2} - \frac{A}{in\pi} \left(e^{\frac{in\pi x}{l}} - e^{\frac{-in\pi x}{l}} + \frac{e^{\frac{in\pi x}{l}}}{3} - \frac{e^{\frac{-in\pi x}{l}}}{3} + \cdots \right),$$

which can be simplified to read

$$f(x) = \frac{A}{2} - \frac{2A}{\pi} \left(\sin \frac{\pi x}{l} + \frac{1}{3} \sin \frac{3\pi x}{l} + \frac{1}{5} \sin \frac{5\pi x}{l} + \cdots \right).$$

This series expansion represents the decomposition of the square-wave function or signal into its various harmonic components.

6.5.2 Summing Numerical Series Using Fourier Series

In some situations, we may need to sum an unfamiliar numerical series. As
pointed out in Section 3.8.5, it may be possible to use the power series
expansion of a function (evaluated at a particular value of x) to sum an
unfamiliar series. Fourier series expansions can similarly be used to sum an
unfamiliar series. As a very quick example, consider summing the numerical
series

$$1 + \frac{1}{3} + \frac{1}{5} + \cdots$$

If we compare this numerical series with the Fourier series expansion of a
square wave found in Section 6.5.1:

$$f(x) = \frac{A}{2} - \frac{2A}{\pi} \left(\sin \frac{\pi x}{l} + \frac{1}{3} \sin \frac{3\pi x}{l} + \frac{1}{5} \sin \frac{5\pi x}{l} + \cdots \right),$$

we can see the similarity between the two series. The terms of the numerical
series are simply the coefficients of the sine terms of the Fourier expansion. To
use the Fourier series expansion to sum our numerical series, we need to find the
point x at which all of the sine functions in the Fourier series evaluate to 1, i.e.,

$$\sin \frac{n\pi x}{l} = 1.$$

Solving for x, we find that

$$x = \frac{l}{2},$$

and evaluating the Fourier series expansion at this point gives us

$$f(l/2) = \frac{A}{2} - \frac{2A}{\pi}\left(1 + \frac{1}{3} + \frac{1}{5} + \cdots\right).$$

Substituting the value for $f(x)$ at $x = l/2$, we obtain

$$A = \frac{A}{2} - \frac{2A}{\pi}\left(1 + \frac{1}{3} + \frac{1}{5} + \cdots\right).$$

This equation can now be simplified to give the sum of the numerical series:

$$\frac{\pi}{4} = \left(1 + \frac{1}{3} + \frac{1}{5} + \cdots\right).$$

6.5.3 Legendre Series Expansion of a Step Function

As an example of a Legendre series expansion, let us find the Legendre series

$$f_1(x) = a_0 P_0 + a_1 P_1 + a_2 P_2 + \cdots$$

of the function $f_1(x)$ shown in Figure 6.2.

The function $f_1(x)$ can be written as

$$f_1(x) = \begin{cases} 1, & -1 < x < 0 \\ 0, & 0 < x < 1 . \end{cases}$$

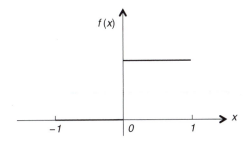

Figure 6.2 A left-facing step function.

To determine the coefficients of the series expansion, we need to evaluate integrals of the form

$$a_l = (l + \frac{1}{2}) \int_{-1}^{1} f(x) P_l(x).$$

For the first few coefficients, we have

$$a_0 = (0 + \frac{1}{2}) \int_{-1}^{1} f(x) P_0(x) dx = \frac{1}{2} \left(\int_{-1}^{0} (1)(1) dx + \int_{0}^{1} (0)(1) dx \right) = \frac{1}{2},$$

$$a_1 = (1 + \frac{1}{2}) \int_{-1}^{1} f(x) P_1(x) dx = \frac{3}{2} \left(\int_{-1}^{0} (1)(x) dx \right) = \frac{-3}{4},$$

$$a_2 = (2 + \frac{1}{2}) \int_{-1}^{1} f(x) P_2(x) dx = \frac{5}{2} \left(\int_{-1}^{0} (1)(\frac{3}{2} x^2 - \frac{1}{2}) dx \right) = 0,$$

and

$$a_3 = (3 + \frac{1}{2}) \int_{-1}^{1} f(x) P_3(x) dx = \frac{7}{2} \left(\int_{-1}^{0} (1)(\frac{5}{2} x^3 - \frac{3}{2} x) dx \right) = \frac{7}{16}.$$

Continuing in this fashion, we obtain the Legendre polynomial expansion

$$f_1(x) = \frac{1}{2} P_0(x) - \frac{3}{4} P_1(x) + \frac{7}{16} P_3(x) + \cdots \tag{6.9}$$

Recall from Section 6.3 that a Legendre series expansion has the property that it converges to the midpoint value at a discontinuity. In this example, there is a discontinuity at $x = 0$. Evaluating our series expression at the origin, we find that

$$f_1(x) = \frac{1}{2} P_0(0) - \frac{3}{4} P_1(0) + \frac{7}{16} P_3(0) + \cdots = \frac{1}{2},$$

confirming that our series expansion converges to the midpoint value at the discontinuity.

What if you were now asked to find the Legendre series expansion for the function shown in Figure 6.3?

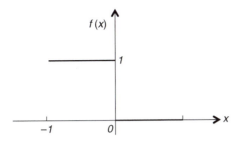

Figure 6.3 A right-facing step function.

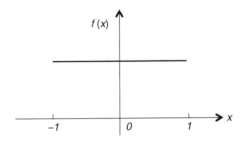

Figure 6.4 The sum of two step functions.

You could work the expansion all over again, but alternatively, a few moments of reflection might lead you to suspect that the series expansion could be given by

$$f_2(x) = \frac{1}{2}P_0(x) + \frac{3}{4}P_1(x) - \frac{7}{16}P_3(x) + \cdots$$

What if you were asked to sum the two functions $f_1(x)$ and $f_2(x)$ shown in Figure 6.4?

What would you suspect the Legendre expansion would look like? It is as follows:

$$f_1(x) + f_2(x) = (\frac{1}{2}P_0(x) - \frac{3}{4}P_1(x) + \frac{7}{16}P_3(x) + \cdots)$$
$$+ (\frac{1}{2}P_0(x) + \frac{3}{4}P_1(x) - \frac{7}{16}P_3(x) + \cdots) = \frac{1}{2}P_0(x) + \frac{1}{2}P_0(x) = 1.$$

Another intriguing property of Legendre series expansions is that they give the best least-squares approximations [8]. In applied problems, we frequently have a given set of data points and would like to find the function $d(x)$ that best fits the data. It is common practice to try and fit the data to a polynomial function of degree n,

$$d(x) = a_0 + a_1 x + a_2 x^2 + \cdots + a_n x^n,$$

and then use the method of least-squares fit to determine the values of the coefficients $a_0, a_1, a_2, \ldots, a_n$ that best fit the data points. For fitting the data on the interval $(-1, 1)$, it can be shown that an expansion in Legendre polynomials will always give the best least-squares fit to $d(x)$.

References

[1] M. F. Ference & H. B. Lemon, *Analytic Experimental Physics*, 4th edn. (Chicago, IL: University of Chicago Press, 1944).

[2] E. W. Swokowski, *Calculus with Analytical Geometry*, 2nd edn. (Boston, MA: Prindel, Weber and Schmidt, 1979).

[3] H. B. Dwight, *Table of Integrals and Other Mathematical Data*, 4th edn. (New York, NY: Macmillan, 1961).

[4] D. Seidman, *The Complete Sailor* (Camden, ME: International Marine, 1995).

[5] G. James, D. Burley, D. Clements, P. Dyke, J. Searl, & J. Wright, *Modern Engineering Mathematics* (Workingham, England: Addison-Wesley, 1992).

[6] R. Gordon & A. Sharov, *Habitability of the Universe before Earth*, Vol. 1 (Cambridge: Academic Press, 2017).

[7] W. Behrens III, D. Meadows, & J. Randers, *The Limits to Growth* (New York, NY: University Books, 1972).

[8] M. Boas, *Mathematical Methods in the Physical Sciences* (Hoboken, NJ: Wiley and Sons, 2006).

[9] K. Knopp, *Theory and Applications of Infinite Series* (Mineola, NY: Dover, 1990).

[10] S. Benson, M. Hobson, & K. Riley, *Mathematical Methods for Physics and Engineering*, 3rd edn. (New York, NY: Cambridge University Press, 2006).

[11] L. Sandler, A Note on the Rule of 72 or How Long It Takes to Double Your Money. *Investment Analysts Journal*, **2**(3) (1973), 34.

[12] Department of the Army, *Field Manual No. 5–23* (Washington, DC: United States Army, 1985).

[13] J. Strong, *Concepts of Classical Optics* (Mineola, NY: Dover, 2004).

[14] D. Borwein, J. Borwein, & K. Taylor, Convergence of Lattice Sums and Madelung's Constant. *Journal of Mathematical Physics*, **26**(1985), 2999–3009.

[15] H. Jessisen, A General Mathematical Form and Description of Contact Angles. *Materialwissenschaft und Werkstofftechnik*, **45**(11) (2014), 961–969.

[16] A. Rabenstein, *Elementary Differential Equations*, 3rd edn. (New York: Academic Press, 1982).

[17] S. Lea, *Mathematics for Physicists* (Belmont, CA: Thomson, Brooks/Cole, 2004).

Index

Printed in the United States
by Baker & Taylor Publisher Services